Vladimir I. Piterbarg

Twenty Lectures
About Gaussian Processes

Atlantic Financial Press

London New York

Professor Vladimir I. Piterbarg
Faculty of Mechanics and Mathematics
Lomonosov Moscow State University
GSP-1 Leninskie Gory
Moscow 119991 Russian Federation
piter@mech.math.msu.su

Mathematics Subject Classification (2010): 60G15, 46G12, 60G55, 60A10, 60D05, 60-01, 60G60

LSI Subjects: MAT029040 (Mathematics : Probability & Statistics – Stochastic Processes), MAT029050 (Mathematics : Probability & Statistics – Time Series), MAT037000 (Mathematics : Functional Analysis)

Suggested reference:

Twenty Lectures About Gaussian Processes, by Vladimir I. Piterbarg, 1st edition, Atlantic Financial Press, 2015

The present volume is the original first softcover edition, first printing.

Errata is available upon request from publisher@atlanticfinancialpress.com

Submit typos and corrections to publisher@atlanticfinancialpress.com

ISBN-10: 0-9844221-9-6
ISBN-13: 978-0-9844221-9-7

Cover art © Anatoly T. Fomenko.
High excursions of trajectories of smooth Gaussian functions are, asymptotically, Morse functions. Methods developed in these lectures could be used to study their asymptotic Euler characteristics. The drawing of a Morse function on the cover, as well as many others by Professor and Academician Anatoly T. Fomenko, can be found at http://dfgm.math.msu.su/files/fomenko/myth-vved.php.

Printed by Lightning Source Ltd.

ref gle.sc.001

Preface

In 1985 I started teaching a course on Gaussian processes to the students of the Faculty of Mechanics and Mathematics of the Lomonosov Moscow State University. The first version of these lectures appeared in print in 1986. Much has changed since then: a half-year course became a full-year one, and the number of lectures increased substantially. Updating the notes has been long overdue, and the latest version was published recently, Питербарг [2015]. Both published editions of the lecture notes were so far only available in Russian, but many non-Russian speakers have told me that they would be interested in the lectures as well. This text is an attempt at transcribing the latest version into English.

This set of lectures consists of two distinct but connected parts. The first part is built around the classic paper Fernique [1975] by Xavier Fernique and covers general theory of Gaussian distributions in vector spaces. The second part is extracted from my monograph Piterbarg [1996] and is devoted to asymptotic methods for Gaussian processes and fields. This part has undergone extensive re-writing from the original "monograph" style to make it much more accessible to students. Here I tried to give only the essence of the results from Piterbarg [1996] while simplifying them as much as possible, particularly trying to avoid technical complications common to "generalized" cases. In particular, as a rule, I only cover Gaussian processes and not fields.

As I have been teaching a course based on these lectures for quite some years now, I am confident that they are reasonably easy to follow by later-year undergraduate and graduate students of our department. I am happy to recommend this set of lectures as a basis for a Ph.D. course elsewhere.

Gaussian processes can be studied without any references to the general theory of random processes, and this is the approach I take. To understand them, it is sufficient to be familiar with the basics of calculus and functional analysis. Of course, those familiar with a wider set of mathematical tools will surely find deeper connections with the theory of Gaussian processes. In particular, some knowledge of the general theory of probability distributions in functional spaces and stochastic processes would make reading easier.

The unique position of the Gaussian distribution in probability theory is due in part to the fact that many calculations can be carried out explicitly, giving us elegant and physically meaningful results. This is particularly useful in applications, from physics to mathematical finance (a good reference here is Andersen and Piterbarg [2010]), where the popularity of the Gaussian model is unrivaled. To keep this text to a reasonable length, I left applications largely outside of its scope, catering more to those who primarily want to learn the underlying tools and methods.

I would like to thank my son Vladimir V. Piterbarg who kindly agreed to be my translation consultant, a (rather strict) editor and a publisher for the English edition. I am also grateful to the students of our faculty: Alexander Zhdanov, Igor Rodionov, Alexander Kleban, Ekaterina Chernavskaya, Julia Gusak and Serik Ajbatov who had the misfortune to endure my lectures and exams, but pointed out many typos and flaws in the text as I was writing it. Sergey Kobelkov helped in organizing this manuscript and I wish to thank him for that. Last but not least, I am delighted that my friend Academician Anatoly T. Fomenko let me use his drawing "A Morse Function and The Euler Characteristic" for the cover. All remaining typos and other problems are solely my own. If you find any, please report them to the publisher at `publisher@atlanticfinancialpress.com`. All those reporting typos will be gratefully acknowledged in the future editions of these lectures.

London, Moscow   *Vladimir I. Piterbarg*

2015

Table of Contents

1

Basic Definitions and Distributions in Finite Dimensions

Let T be an arbitrary set, and $(\Omega, \mathcal{F}, \mathrm{P})$ be a probability space. A family of random variables $\mathbf{X} = \{X_t(\omega), t \in T, \omega \in \Omega\}$ is called a *real-valued Gaussian function indexed by the parametric set T* if, for any finite subset $T_0 \subset T$, the random vector $\mathbf{X}(T_0) = (X_t(\omega), t \in T_0\}$ taking values in $\mathbb{R}^{T_0}$, is Gaussian. If $T \subset \mathbb{R}$, $\mathbf{X}$ is called a Gaussian random process, and if $T \subset \mathbb{R}^n$, it is called a Gaussian random field. For a fixed ω, the real function $X(\cdot, \omega)$ is called a trajectory, or sample path, of the Gaussian random function. In this lecture we define Gaussian finite-dimensional vectors and consider their main properties.

1.1 Definitions and Basic Properties

Definition 1.1. *A random variable $X = X(\omega)$ taking values in $(\mathbb{R}, \mathcal{B})$, where $\mathcal{B}$ is a Borel σ-algebra, is called* Gaussian *(or* Normal*) if its characteristic function is given by*

$$\psi(z) := \mathrm{E}e^{izX} = \exp\left(izm - \frac{1}{2}\sigma^2 z^2\right).$$

From now on we assume that $\sigma \geqslant 0$.

Differentiating $\psi(z)$, we obtain that $\mathrm{E}X = m$ and $\mathrm{Var}X = \sigma^2$. For a Gaussian X we use the notation $X \sim \mathcal{N}(m, \sigma^2)$. If $\sigma = 0$ then $X = m$ almost surely (a.s.), so in this case X is called a degenerate Normal variable. A random variable $X \sim \mathcal{N}(0, 1)$ is called a standard Normal or standard Gaussian.

Definition 1.2. *A random vector[1] $\mathbf{X} = \mathbf{X}(\omega) = (X_1, \ldots, X_d)^\top$ taking values in a measurable Euclidean space $(\mathbb{R}^d, \mathcal{B}^d)$, $\mathcal{B}^d$ is a Borel σ-algebra, is called Gaussian, if its characteristic function*

---

[1] Vectors are defined to be column vectors.

$$\psi(\mathbf{z}) = \mathrm{E}e^{i(\mathbf{z},\mathbf{X})}, \ \mathbf{z} = (z_1,\dots,z_d)^\top,$$

equals

$$\psi(\mathbf{z}) = \exp\left(i(\mathbf{z},\mathbf{m}) - \frac{1}{2}(\mathbf{z},R\mathbf{z})\right), \tag{1.1}$$

with $\mathbf{m}$, *a vector, and* R, *a matrix.*

As $\psi(\mathbf{z})$ must be bounded, the matrix R must be non-negative definite. The vector $\mathbf{m} = (m_1,\dots,m_d)^\top$ is the vector of the expected values of the components of $\mathbf{X} = (X_1,\dots,X_d)^\top$; the matrix $R = (r_{kl} : k,l = 1,\dots,d)$ is the matrix of its covariances. This can be seen by differentiating the characteristic function at zero,

$$m_k = \mathrm{E}X_k = -i\frac{\partial\psi(\mathbf{z})}{\partial z_k}\Big|_{z_k=0}, \ \text{and } \mathrm{E}X_kX_l = r_{kl} + m_k m_l = -\frac{\partial^2\psi(\mathbf{z})}{\partial z_k \partial z_l}\Big|_{z_k=z_l=0}.$$

Now let us give an equivalent definition of a Gaussian vector that, in particular, shows that $\psi(\mathbf{z})$ is indeed a characteristic function.

Definition 1.3. *A random vector* $\mathbf{X}$ *is called a standard Gaussian, or standard Normal, if its components* $X_1,\dots,X_d$ *are independent and for every i,* $X_i \sim \mathcal{N}(0,1)$. *A random vector* $\mathbf{X}$ *is called Gaussian if, for some vector* $\mathbf{m}$, *a matrix* A, *and a standard Gaussian vector* $\mathbf{Z}$,

$$\mathbf{X} \overset{d}{=} \mathbf{m} + A\mathbf{Z}.$$

The notation $\overset{d}{=}$ *means equality in distribution.*

In order to show the equivalence of the two definitions 1.2 and 1.3, let us compute the characteristic function of a Gaussian vector as defined by the second definition. Notice that the characteristic function of a Gaussian standard vector $\mathbf{Z}$ equals

$$\prod_{i=1}^{d} \exp(-z_i^2/2) = \exp(-|\mathbf{z}|^2/2).$$

Furthermore,

$$\mathrm{E}\exp\left(i(\mathbf{z},\mathbf{m} + A\mathbf{Z})\right) = e^{i(\mathbf{z},\mathbf{m})}\mathrm{E}\exp\left(\mathbf{z},A\mathbf{Z}\right)$$

$$= \exp\left(i(\mathbf{z},\mathbf{m})\right)\mathrm{E}\exp\left(A^\top\mathbf{z},\mathbf{Z}\right)$$

$$= \exp(i(\mathbf{z},\mathbf{m}))\exp\left(-\frac{1}{2}|A^\top\mathbf{z}|^2\right)$$

$$= \exp\left(i(\mathbf{z},\mathbf{m}) - \frac{1}{2}(\mathbf{z},R\mathbf{z})\right),$$

where $R = AA^\top$, so that R is non-negative definite. Thus, the first definition follows from the second. Conversely, let $\mathbf{m}$ and R be the parameters of a

Gaussian vector in the sense of the definition 1.2. Let A be a square root of R so that $R = AA^\top$. Let $\mathbf{Z}$ be a standard Gaussian vector. Then the characteristic function of the vector $\mathbf{X} = \mathbf{m} + A\mathbf{Z}$ is equal to what is given in Definition 1.2. Therefore, the definitions are equivalent, and $\psi(\mathbf{z})$ is indeed a valid characteristic function.

Let $\mathbf{X} = (\mathbf{X}_1, \mathbf{X}_2)$ be a Gaussian vector, where its components $\mathbf{X}_1$ and $\mathbf{X}_2$ can also be vectors. The covariance matrix of $\mathbf{X}$ has the following structure,

$$R = \begin{pmatrix} R_{11} & R_{12} \\ R_{21} & R_{22} \end{pmatrix},$$

where R_{11} and R_{22} are the covariance matrices of $\mathbf{X}_1$ and $\mathbf{X}_2$ respectively, and the rectangular matrix R_{12} is called the *matrix of cross covariances*. By the symmetry of R, $R_{12} = R_{21}^\top$, the latter being the conjugate matrix to R_{21}.

The characteristic function of $\mathbf{X}$ can be represented as

$$\psi_{\mathbf{X}}(\mathbf{z}) = \exp(i(\mathbf{z}_1, \mathbf{m}_1) + i(\mathbf{z}_2, \mathbf{m}_2) - \frac{1}{2}(\mathbf{z}_1, R_{11}\mathbf{z}_1) - \frac{1}{2}(\mathbf{z}_2, R_{22}\mathbf{z}_2)$$
$$- \frac{1}{2}(\mathbf{z}_1, R_{12}\mathbf{z}_2) - \frac{1}{2}(\mathbf{z}_2, R_{21}\mathbf{z}_1)), \qquad (1.2)$$

where $(\mathbf{m}_1, \mathbf{m}_2) = E\mathbf{X}$, and $\mathbf{z}_1, \mathbf{z}_2$ are the sub-vector arguments that correspond to $\mathbf{X}_1, \mathbf{X}_2$, respectively. From the representation (1.2) it follows that the sub-vectors $\mathbf{X}_1$ and $\mathbf{X}_2$ are independent if and only if $R_{12} = 0$, that is, all of the cross-covariances of $\mathbf{X}_1$ and $\mathbf{X}_2$ are equal to zero. Vectors like this are called uncorrelated. Henceforth, the following holds.

Proposition 1.4. *Let the Gaussian random vectors $\mathbf{X}_1$ and $\mathbf{X}_2$ be such that the compound vector $(\mathbf{X}_1, \mathbf{X}_2)$ is also Gaussian. Then $\mathbf{X}_1$ and $\mathbf{X}_2$ are independent if and only if they are uncorrelated.*

Problem 1.5. Give an example of two Gaussian random variables that are uncorrelated but dependent.

Hint: Let X_1, Y_1 be independent standard Gaussian variables. One can set

$$(X, Y) = \begin{cases} (X_1, |Y_1|), & \text{if } X_1 \geqslant 0, \\ (X_1, -|Y_1|), & \text{if } X_1 < 0. \end{cases}$$

Problem 1.6. Give an example of two Gaussian random variables with a non-Gaussian joint distribution.

Hint: See the previous problem. Try to give another example.

Proposition 1.7. *The sum of two independent Gaussian vectors is Gaussian.*

Proof. Since the characteristic function of a sum of independent random vectors equals the product of their characteristic functions, we have,

$$\psi_{\mathbf{X}_1+\mathbf{X}_2}(\mathbf{z}) = \exp(i\mathbf{z}(\mathbf{m}_1 + \mathbf{m}_2) - \frac{1}{2}(\mathbf{z}, (R_1 + R_2)\mathbf{z}),$$

where $(\mathbf{m}_1, R_1)$ and $(\mathbf{m}_2, R_2)$ are the respective parameters of $\mathbf{X}_1$ and $\mathbf{X}_2$.

Let us derive the density distribution of a Gaussian vector when one exists. First, consider the one-dimensional case and find the density of a Gaussian standard random variable. Its characteristic function is summable; hence we can find its Fourier inversion,

$$\varphi(x) = \frac{1}{2\pi} \int e^{-izx - z^2/2} dz = \frac{1}{2\pi} e^{-x^2/2} \int e^{-(z+ix)^2/2} dz$$

$$= \frac{1}{2\pi} e^{-x^2/2} \int_{ix-\infty}^{ix+\infty} e^{-z^2/2} dz = \frac{1}{2\pi} e^{-x^2/2} \int_{-\infty}^{\infty} e^{-z^2/2} dz.$$

Note that in texts in probability theory the integral sign $\int$ is understood, as a rule, as $\int_{-\infty}^{+\infty}$. The last equality holds because $e^{-z^2/2}$ is an analytical function. The integral can be calculated as following,

$$\int e^{-z^2/2} dz = \left(\int\int e^{-u^2/2 + v^2/2} du dv \right)^{1/2},$$

and, using polar coordinates,

$$\int\int e^{-u^2/2 + v^2/2} du dv = 2\pi \int_0^{\infty} r e^{-r^2/2} dr = 2\pi \int_0^{\infty} e^{-s} ds = 2\pi.$$

Thus,

$$\varphi(x) = \frac{1}{\sqrt{2\pi}} e^{-x^2/2}. \tag{1.3}$$

From this we get immediately that the density of the standard Gaussian vector exists and is equal to

$$\varphi(\mathbf{x}) = \frac{1}{(2\pi)^{d/2}} \exp\left(-\frac{1}{2}||\mathbf{x}||^2\right). \tag{1.4}$$

Proposition 1.8. *Let* $\mathbf{X}$ *be a Gaussian random vector with the parameters* $(\mathbf{m}, R)$, *and let its covariance matrix R be non-degenerate, that is,* $\mathrm{rank}(R) = d$. *Then its density exists and is equal to*

$$\varphi_{\mathbf{X}}(\mathbf{x}) = \frac{1}{(2\pi)^{d/2}(\det R)^{1/2}} \exp\left(-\frac{1}{2}(\mathbf{x} - \mathbf{m}, R^{-1}(\mathbf{x} - \mathbf{m}))\right), \tag{1.5}$$

where det *means the determinant of a matrix.*

Proof. By Definition 1.3, $\mathbf{X} \overset{d}{=} \mathbf{m} + A\mathbf{Z}$, where A is non-degenerate. Hence $R = AA^{\top}$ is also non-degenerate. The density of $\mathbf{Z}$ is given by the standard transform rule for linear transforms of a random vector, that is,

$$\varphi_{A\mathbf{Z}+\mathbf{m}}(x) = (\det A)^{-1} \varphi_{\mathbf{Z}}(A^{-1}(\mathbf{x} - \mathbf{m})).$$

Applying this to the density of a standard Gaussian vector $\mathbf{Z}$ and taking into account that $\det R = (\det A)^2$, we get the claimed representation of the density.

The formula (1.5) can also be obtained by the inverse Fourier transform of (1.1),

$$\varphi_{\mathbf{X}}(\mathbf{x}) = (2\pi)^{-d} \int_{\mathbb{R}^d} e^{-i(\mathbf{x},\mathbf{z}) - \frac{1}{2}(\mathbf{z}, R\mathbf{z})} d\mathbf{z}.$$

Differentiating this we get useful relations for a Gaussian density,

$$\frac{\partial^2 \varphi_{\mathbf{X}}(\mathbf{x})}{\partial x_k \partial x_l} = \frac{\partial \varphi_{\mathbf{X}}(\mathbf{x})}{\partial r_{kl}}, \ k \neq l, \text{ and } \frac{1}{2}\frac{\partial^2 \varphi_{\mathbf{X}}(\mathbf{x})}{\partial x_k^2} = \frac{\partial \varphi_{\mathbf{X}}(\mathbf{x})}{\partial r_{kk}}, \ k,l = 1,\ldots,d. \quad (1.6)$$

It means that the Gaussian density is a solution to the so-called *heat equation*.

Now let us consider the case of a degenerate covariance matrix R of a Gaussian vector $\mathbf{X}$ with $\text{rank}(R) = r < d$. Note that then $\text{rank}(A) = r$ as well. The image of the Euclidean space $\mathbb{R}^d$ under the transform $\mathbf{x} \mapsto A\mathbf{x} + \mathbf{m}$ is an r-dimensional hyperplane in $\mathbb{R}^d$. Let us denote it by $\mathbb{L}$. Then,

$$P(\mathbf{X} \in \mathbb{L}) = 1.$$

Suppose now that there exists a set $D \subset \mathbb{L}$ with non-zero r-volume and $P(\mathbf{X} \in D) = 0$. Then, for the set $D' := \{\mathbf{x} : A\mathbf{x} + \mathbf{m} \in D\}$ with non-zero volume in $\mathbb{R}^d$, $P(\mathbf{Z} \in D') = 0$, which is impossible for a standard Gaussian vector, as its density is positive everywhere. Thus we have proven the following.

Proposition 1.9. *Let $\mathbf{X}$ be a Gaussian random vector with the parameters $(\mathbf{m}, R)$, and let $\text{rank}(R) = r < d$. Then there exists a hyperplane in $\mathbb{R}^d$ with dimension r which is the support of the distribution of $\mathbf{X}$. This hyperplane is the image of $\mathbb{R}^d$ under the transform $\mathbf{x} \mapsto A\mathbf{x} + \mathbf{m}$, where A is the square root of R.*

1.2 Conditional Gaussian Distributions

Since by convention all our vectors are column vectors, the matrix of cross-covariances between two random vectors $\mathbf{X}$ and $\mathbf{Y}$, even if they are of different dimensions, can be written as $E(\mathbf{X}-E\mathbf{X})(\mathbf{Y}-E\mathbf{Y})^{\top}$. In particular, the covariance matrix of a random vector $\mathbf{X}$ is given by $E(\mathbf{X}-E\mathbf{X})(\mathbf{X}-E\mathbf{X})^{\top}$. Many results on conditional distributions of Gaussian variables are based on the following theorem.

Theorem 1.10. *Let* $(\mathbf{X}, \mathbf{Y})$ *be a Gaussian random vector, where* $\mathbf{X}$ *and* $\mathbf{Y}$ *can also be vectors. Let the distribution of* $\mathbf{Y}$ *be non-degenerate. We define the following matrix,*

$$\rho(\mathbf{X} \,|\, \mathbf{Y}) = \mathrm{E}(\mathbf{X} - \mathrm{E}\mathbf{X})(\mathbf{Y} - \mathrm{E}\mathbf{Y})^\top \left(\mathrm{E}(\mathbf{Y} - \mathrm{E}\mathbf{Y})(\mathbf{Y} - \mathrm{E}\mathbf{Y})^\top \right)^{-1}.$$

Then $\mathbf{Y}$ *and* $\mathbf{X} - \rho(\mathbf{X} \,|\, \mathbf{Y})\mathbf{Y}$ *are independent. Moreover,*

$$\mathrm{E}(\mathbf{X} \,|\, \mathbf{Y}) = \mathrm{E}\mathbf{X} + \rho(\mathbf{X} \,|\, \mathbf{Y})(\mathbf{Y} - \mathrm{E}\mathbf{Y}). \tag{1.7}$$

The matrix $\rho(\mathbf{X} \,|\, \mathbf{Y})$ is a generalization of the coefficient of regression of a one-dimensional $\mathbf{X}$ against $\mathbf{Y}$. It is sometimes known as a *matrix regression coefficient*.

Proof. The independence of $\mathbf{Y}$ and $\mathbf{X} - \rho(\mathbf{X} \,|\, \mathbf{Y})\mathbf{Y}$ is equivalent to independence of $\mathbf{Y} - \mathrm{E}\mathbf{Y}$ and $\mathbf{X} - \mathrm{E}\mathbf{X} - \rho(\mathbf{X} \,|\, \mathbf{Y})(\mathbf{Y} - \mathrm{E}\mathbf{Y})$, hence we can assume that $\mathrm{E}\mathbf{X} = 0$ and $\mathrm{E}\mathbf{Y} = 0$. Furthermore,

$$
\begin{aligned}
\mathrm{E}\mathbf{Y}(\mathbf{X} - \rho(\mathbf{X} \,|\, \mathbf{Y})\mathbf{Y})^\top &= \mathrm{E}\mathbf{Y}\mathbf{X}^\top - \mathrm{E}\mathbf{Y}\mathbf{Y}^\top \rho(\mathbf{X} \,|\, \mathbf{Y})^\top \\
&= \mathrm{E}\mathbf{Y}\mathbf{X}^\top - \mathrm{E}\mathbf{Y}\mathbf{Y}^\top \left(\mathrm{E}\mathbf{Y}\mathbf{Y}^\top \right)^{-1} \left(\mathrm{E}\mathbf{X}\mathbf{Y}^\top \right)^\top \\
&= \mathrm{E}\mathbf{Y}\mathbf{X}^\top - \mathrm{E}\mathbf{Y}\mathbf{X}^\top = 0,
\end{aligned}
$$

and by Proposition 1.4, the vectors $\mathbf{Y}$ and $\mathbf{X} - \rho(\mathbf{X} \,|\, \mathbf{Y})\mathbf{Y}$ are independent. Taking the conditional expected value of both sides of the equality

$$\mathbf{X} = \rho\mathbf{Y} + (\mathbf{X} - \rho\mathbf{Y}),$$

leads to (1.7). $\quad\square$

Corollary 1.11. *Consider a Gaussian vector* $(\mathbf{X}, \mathbf{Y})$ *with possibly vector components* $\mathbf{X}$ *and* $\mathbf{Y}$. *The conditional expected value of* $\mathbf{X}$ *given* $\mathbf{Y}$ *is linear with respect to the condition and can be calculated by (1.7). The matrix of conditional covariances*

$$\mathrm{E}\left((\mathbf{X} - \mathrm{E}(\mathbf{X}|\mathbf{Y}))(\mathbf{X} - \mathrm{E}(\mathbf{X}|\mathbf{Y}))^\top | \mathbf{Y} \right)$$

is non-random and equals $\mathrm{E}\mathbf{Z}\mathbf{Z}^\top$, *where* $\mathbf{Z} = \mathbf{X} - \mathrm{E}(\mathbf{X} \,|\, \mathbf{Y})$.

The first part of the corollary is already proven. The second part follows from the independence of $\mathbf{Y}$ and $\mathbf{Z}$.

Since $\mathbf{Y}$ and $\mathbf{Z}$ are independent, the covariance matrix of $\mathbf{X}$ equals the sum of the covariance matrices of the vectors $\rho\mathbf{Y}$ and $\mathbf{Z}$, and we obtain the following result.

Corollary 1.12. *The matrix*

$$\mathrm{E}(\mathbf{X} - \mathrm{E}\mathbf{X})(\mathbf{X} - \mathrm{E}\mathbf{X})^\top - \mathrm{E}((\mathbf{X} - \mathrm{E}(\mathbf{X}|\mathbf{Y}))(\mathbf{X} - \mathrm{E}(\mathbf{X}|\mathbf{Y})^\top | \mathbf{Y})$$

is non-negative-definite. In particular, in case $\dim(\mathbf{X}) = 1$ *we have* $\mathrm{Var}X \geqslant \mathrm{Var}(X|\mathbf{Y})$: *the conditional variance is always less than or equal to the unconditional one.*

1.3 Useful Formulas

Those who work with Gaussian multi-dimensional distributions often need to compute conditional means, variances and covariances. We list some useful formulas in this section; they are easily derived from the results above. The following results mostly come from Anderson [2003] with small changes to accommodate our notations. Here $(\mathbf{X}, \mathbf{Y})$ is a Gaussian vector, where $\mathbf{X}$ and $\mathbf{Y}$ can also be vectors. We have

$$R_{\mathbf{XX}} = \mathrm{E}(\mathbf{X}-\mathrm{E}\mathbf{X})(\mathbf{X}-\mathrm{E}\mathbf{X})^\top, \ R_{\mathbf{YY}} = \mathrm{E}(\mathbf{Y}-\mathrm{E}\mathbf{Y})(\mathbf{Y}-\mathrm{E}\mathbf{Y})^\top,$$

$$R_{\mathbf{XY}} = \mathrm{E}(\mathbf{X}-\mathrm{E}\mathbf{X})(\mathbf{Y}-\mathrm{E}\mathbf{Y})^\top, \ R_{\mathbf{YX}} = \mathrm{E}(\mathbf{Y}-\mathrm{E}\mathbf{Y})(\mathbf{X}-\mathrm{E}\mathbf{X})^\top,$$

$$R_{\mathbf{XX}|\mathbf{Y}} = \mathrm{E}\left((\mathbf{X} - \mathrm{E}(\mathbf{X}|\mathbf{Y}))(\mathbf{X} - \mathrm{E}(\mathbf{X}|\mathbf{Y}))^\top | \mathbf{Y}\right),$$

denoting the matrices of covariances, cross-covariances and conditional covariances between the two vectors. Assuming again that $R_{\mathbf{YY}}$ is non-degenerate, we obtain that

$$R_{\mathbf{XX}|\mathbf{Y}} = R_{\mathbf{XX}} - R_{\mathbf{XY}} R_{\mathbf{YY}}^{-1} R_{\mathbf{YX}}$$

and

$$\mathrm{E}(\mathbf{X}|\mathbf{Y}) = \mathrm{E}\mathbf{X} + R_{\mathbf{XY}} R_{\mathbf{YY}}^{-1}(\mathbf{Y} - \mathrm{E}\mathbf{Y}),$$

or, specifying $\mathbf{Y}$,

$$\mathrm{E}(\mathbf{X}|\mathbf{Y} = \mathbf{y}) = \mathrm{E}\mathbf{X} + R_{\mathbf{XY}} R_{\mathbf{YY}}^{-1}(\mathbf{y} - \mathrm{E}\mathbf{Y}).$$

The matrix $\rho(\mathbf{X}\,|\,\mathbf{Y}) = R_{\mathbf{XY}} R_{\mathbf{YY}}^{-1}$ is also known as the matrix of regression coefficients of $\mathbf{X}$ regressed against $\mathbf{Y} = \mathbf{y}$.

2

Comparison of Distributions in Finite Dimensions

As a start, let us recall an important property of a Gaussian density that we established in the first lecture. Recall that the distribution of a Gaussian random vector $\mathbf{X}$ has a density if and only if its characteristic function is summable, that is, its covariance matrix is positive-definite (we often simply call such matrices *positive*). Let this be the case, and suppose for simplicity that the expected value of the Gaussian vector equals zero, that is, $\mathbf{m} = \mathbf{0}$. Taking an inverse Fourier transform we obtain,

$$\varphi(\mathbf{x}) = (2\pi)^{-d} \int_{\mathbb{R}^d} e^{-i(\mathbf{z},\mathbf{x}) - \frac{1}{2}(\mathbf{z}, R\mathbf{z})} d\mathbf{z}.$$

Differentiating, we obtain

$$\frac{\partial^2 \varphi}{\partial x_i \partial x_j} = \frac{\partial \varphi}{\partial r_{ij}}, \ i \neq j, \ \text{ and } \ \frac{1}{2}\frac{\partial^2 \varphi}{\partial x_i^2} = \frac{\partial \varphi}{\partial r_{ii}}, \tag{2.1}$$

where r_{ij} are elements of R. This property of a Gaussian density is well-known and is quite useful when working with Gaussian distributions, as we shall see in this lecture. The function φ is alternatively called either a *fundamental solution* of the heat equation, or the Poisson kernel. The first important property of a finite-dimensional Gaussian distribution follows from (2.1):

Lemma 2.1. *The distribution function of a Gaussian vector is an increasing function of its covariances (i.e. the off-diagonal elements of the covariance matrix).*

Proof. Let us first assume that the distribution function in question $F(x_1, \ldots, x_d)$ has a density, that is,

$$F(x_1, \ldots, x_d) = \int_{-\infty}^{x_1} \cdots \int_{-\infty}^{x_d} \varphi(y_1, \ldots, y_d) d\mathbf{y}.$$

Differentiating and then integrating, we obtain,

$$\frac{\partial F}{\partial r_{ij}}$$

$$= \int_{-\infty}^{x_1} \cdots \int_{-\infty}^{x_d} \frac{\partial}{\partial r_{ij}} \varphi(y_1,\ldots,y_d) d\mathbf{y}$$

$$= \int_{-\infty}^{x_1} \cdots \int_{-\infty}^{x_d} \frac{\partial^2}{\partial y_i \partial y_j} \varphi(y_1,\ldots,y_d) d\mathbf{y}$$

$$= \int_{-\infty}^{x_1} \cdots \int_{-\infty}^{x_d} \varphi(y_1,\ldots,y_{i-1},x_i,y_{i+1},\ldots,y_{j-1},x_j,y_{j+1},\ldots,y_d) \frac{d\mathbf{y}}{dx_i dx_j}$$

$$\geqslant 0,$$

where the multiplicity of the last integral is $d - 2$, and the formal ratio denotes the corresponding $(d - 2)$-fold differential.

Now, in case the density distribution of $\mathbf{X}$ does not exist, we would consider a Gaussian vector $\mathbf{X} + \varepsilon \Lambda$, where $\varepsilon > 0$ and Λ is the standard Gaussian vector independent of $\mathbf{X}$, that is, a vector with independent standard Gaussian components which also do not depend of $\mathbf{X}$. Covariances of this vector are still equal to r_{ij}, $i \neq j$, it does have a density, and so its distribution function increases in r_{ij}. Letting ε go to zero establishes the required property of F.

2.1 Slepian's Inequality

Other useful relations between Gaussian distributions could be derived from (2.1). Let $f : \mathbb{R}^d \mapsto \mathbb{R}$ be a measurable function, to be more fully specified later. We consider two Gaussian vectors , $\mathbf{X}_0$ and $\mathbf{X}_1$, and assume for now that they are independent (although we shall see that this is not a restriction for the result we are deriving). Consider a family of Gaussian vectors, $\mathbf{X}_h = \sqrt{h}\mathbf{X}_1 + \sqrt{1-h}\mathbf{X}_0$, $0 \leqslant h \leqslant 1$. The covariance matrix of the vector $\mathbf{X}_h$ is equal, by independence, to $R_h = hR_1 + (1 - h)R_0$, where R_0 and R_1 are the covariance matrices of the original vectors. Let us denote the elements of this matrix by r_{hij}. By symmetry of R_h, we formally have that

$$\mathrm{E}f(\mathbf{X}_1) - \mathrm{E}f(\mathbf{X}_0) = \int_0^1 \frac{d}{dh} \mathrm{E}f(\mathbf{X}_h) dh = \int_0^1 \sum_{i \geqslant j=1}^{d} \frac{\partial}{\partial r_{hij}} \mathrm{E}f(\mathbf{X}_h) \frac{dr_{hij}}{dh} dh$$

$$= \sum_{i \geqslant j=1}^{d} (r_{1ij} - r_{0ij}) \int_0^1 \frac{\partial}{\partial r_{hij}} \mathrm{E}f(\mathbf{X}_h) dh.$$

Now let us justify the ability to differentiate and the ability to change the order of integration in the last equality. More precisely, let us actually state the conditions under which these operations can be performed. First let us assume that the distributions of $\mathbf{X}_0$ and $\mathbf{X}_1$ have densities; the same is then true for any vector $\mathbf{X}_h$. Let us denote its density by φ_h. Furthermore, let us assume that f is twice differentiable, and that for any $\varepsilon > 0$,

$$\limsup_{\mathbf{x} \to \infty}(|f(\mathbf{x})| + ||\nabla f(\mathbf{x})|| + ||\nabla\nabla^\top f(\mathbf{x})||)e^{-\varepsilon||\mathbf{x}||^2} < \infty, \qquad (2.2)$$

where $||\nabla\nabla^\top f(\mathbf{x})||$ is the Euclidean (Frobenius) norm of the Hessian matrix of f, i.e. the matrix of its second derivatives. Under these restrictions, all the formal steps above are justified and, moreover, for any $i \neq j$, we have,

$$\frac{\partial}{\partial r_{hij}}\mathrm{E}f(\mathbf{X}_h) = \int_{\mathbb{R}^d} \frac{\partial}{\partial r_{hij}}f(\mathbf{x})\varphi_h(\mathbf{x})d\mathbf{x} = \int_{\mathbb{R}^d} f(\mathbf{x})\frac{\partial^2}{\partial x_i \partial x_j}\varphi_h(\mathbf{x})d\mathbf{x}$$

$$= \int_{\mathbb{R}^d} \varphi_h(\mathbf{x})\frac{\partial^2 f(\mathbf{x})}{\partial x_i \partial x_j}d\mathbf{x} = \mathrm{E}\frac{\partial^2}{\partial x_i \partial x_j}f(\mathbf{X}_h).$$

For $i = j$ the argument is a bit different,

$$\frac{\partial}{\partial r_{hii}}\mathrm{E}f(\mathbf{X}_h) = \int_{\mathbb{R}^d} \frac{\partial}{\partial r_{hii}}f(\mathbf{x})\varphi_h(\mathbf{x})d\mathbf{x} = \frac{1}{2}\int_{\mathbb{R}^d} f(\mathbf{x})\frac{\partial^2}{\partial x_i^2}\varphi_h(\mathbf{x})d\mathbf{x}$$

$$= \frac{1}{2}\int_{\mathbb{R}^d} \frac{\partial^2}{\partial x_i^2}f(\mathbf{x})\varphi_h(\mathbf{x})d\mathbf{x} = \frac{1}{2}\mathrm{E}\frac{\partial^2}{\partial x_i^2}f(\mathbf{X}_h).$$

Thus we obtain that

$$\mathrm{E}f(\mathbf{X}_1) - \mathrm{E}f(\mathbf{X}_0) = \frac{1}{2}\sum_{i,j=1}^{d}(r_{1ij} - r_{0ij})\int_0^1 \mathrm{E}\frac{\partial^2}{\partial x_i \partial x_j}f(\mathbf{X}_h)dh. \qquad (2.3)$$

This identity can be used in several ways. For instance, suppose that all first- and second-order derivatives of f are non-negative. For example, if $f(\mathbf{x}) = \prod_{i=1}^{d} f_i(x_i)$ satisfies these conditions, then all f_k are non-negative, monotone and concave. If $r_{1ij} \geq r_{0ij}$ for all i,j, then $\mathrm{E}f(\mathbf{X}_1) \geq \mathrm{E}f(\mathbf{X}_0)$.

Another result that we can obtain here is the celebrated *Slepian inequality*. Suppose that $r_{1ij} \geq r_{0ij}$ for all $i \neq j$, but $r_{ii1} \equiv r_{ii0}$. Let also $f(\mathbf{x}) = \prod_{i=1}^{d} f_i(x_i)$, and all f_i are monotone, and either all non-increasing or non-decreasing. Then again, $\mathrm{E}f(\mathbf{X}_1) \geq \mathrm{E}f(\mathbf{X}_0)$. Moreover, an indicator function $\mathbf{I}_{(-\infty,x)}$ can be approximated with non-increasing smooth functions, leading to Slepian's inequality.

Lemma 2.2 (Slepian's Inequality). *Let $\mathbf{X}_1$ and $\mathbf{X}_0$ be two Gaussian vectors in $\mathbb{R}^d$, with zero mean and the covariance matrices r_{1ij} and r_{0ij}, respectively. If $r_{1ij} \geq r_{0ij}$ for all $i \neq j$, while $r_{1ii} \equiv r_{0ii}$, then for any $\mathbf{x} = (x_1,...,x_d)$,*

$$\mathrm{P}(X_{11} \leq x_1,...,X_{1d} \leq x_d) \geq \mathrm{P}(X_{01} \leq x_1,...,X_{0d} \leq x_d).$$

That is, deceasing the dependencies between the components of a Gaussian vector, while keeping their variances the same, makes its distribution function lower.

2.2 Fernique-Sudakov's Inequality

Another important comparison result is the Fernique-Sudakov's inequality.

Lemma 2.3. *Let $\mathbf{X}_1$, $\mathbf{X}_0$ be two Gaussian random vectors in $\mathbb{R}^d$ with zero means and covariance matrices r_{1ij} and r_{0ij}, respectively. Denote*

$$d_v^2(i,j) = \mathrm{E}(X_{vj} - X_{vi})^2, \ v = 0,1, \ i,j = 1,\ldots,d$$

and

$$\delta := \max_{i,j=1,\ldots,d} |d_1^2(i,j) - d_0^2(i,j)|.$$

Then

$$|\mathrm{E}\max_i X_{0i} - \mathrm{E}\max_i X_{1i}| \leq \sqrt{2\delta \ln d}. \qquad (2.4)$$

If, moreover, $d_0^2(i,j) \leq d_1^2(i,j)$ for all $i,j = 1,\ldots,d$, then

$$\mathrm{E}\max_i X_{0i} \leq \mathrm{E}\max_i X_{1i}. \qquad (2.5)$$

Proof. Let us introduce a function

$$g_\beta(\mathbf{x}) = \beta^{-1}\ln\sum_{i=1}^d e^{\beta x_i}.$$

We notice that $g_\beta(\mathbf{x}) \to \max_i x_i$ as $\beta \to \infty$, and (2.2) holds for g_β. Furthermore,

$$\max_i x_i = \beta^{-1}\ln\frac{1}{d}\sum_{j=1}^d e^{\beta\max_i x_i} \leq g_\beta(\mathbf{x}) \leq \beta^{-1}\ln(de^{\beta\max_i x_i})$$

$$\leq \beta^{-1}\ln d + \max_i x_i,$$

hence

$$\sup_{\mathbf{x}} |g_\beta(\mathbf{x}) - \max_i x_i| \leq \beta^{-1}\ln d. \qquad (2.6)$$

We can now apply the comparison identity (2.3). Notice that

$$\frac{\partial g_\beta}{\partial x_i}(\mathbf{x}) = \frac{e^{\beta x_i}}{\sum_{j=1}^d e^{\beta x_j}} =: p_i(\mathbf{x}).$$

It is easy to check that

$$\frac{\partial^2 g_\beta}{\partial x_i \partial x_j}(\mathbf{x}) = \begin{cases} \beta(p_i(\mathbf{x}) - p_i^2(\mathbf{x})), & i = j; \\ -\beta p_i(\mathbf{x})p_j(\mathbf{x}), & i \neq j. \end{cases}$$

Since $\sum p_i(\mathbf{x}) = 1$, we get that

$$\sum_{i,j=1}^{d}(r_{1ij}-r_{0ij})\frac{\partial^2 g_\beta}{\partial x_i \partial x_j}(\mathbf{x})$$

$$= \beta \sum_{i=1}^{d}(r_{1ii}-r_{0ii})p_i(\mathbf{x}) - \beta \sum_{i,j=1}^{d}(r_{1ij}-r_{0ij})p_i(\mathbf{x})p_j(\mathbf{x})$$

$$= \frac{\beta}{2}\sum_{ij=1}^{d}[(r_{1ii}+r_{1jj}-2r_{1ij})-(r_{0ii}+r_{0jj}-2r_{0ij})]p_i(\mathbf{x})p_j(\mathbf{x})$$

$$= \frac{\beta}{2}\sum_{i,j=1}^{d}(d_1^2(i,j)-d_0^2(i,j))p_i(\mathbf{x})p_j(\mathbf{x}) \geqslant 0.$$

Under the conditions of the lemma this inequality, together with the identity (2.3), holds for all positive β. Thus, letting β go to infinity, we obtain (2.5).

We can now find an upper bound for the right-hand part of (2.3). Specifically, we have,

$$\left|\sum_{ij=1}^{d}(r_{1ij}-r_{0ij})\frac{\partial^2 g_\beta}{\partial x_i \partial x_j}(\mathbf{x})\right| \leqslant \frac{\beta\delta}{2}\sum_{ij=1}^{d}p_i(\mathbf{x})p_j(\mathbf{x}) = \frac{\beta\delta}{2},$$

so that

$$|E g_\beta(\mathbf{X}_1) - E g_\beta(\mathbf{X}_0)| \leqslant \frac{\beta\delta}{4}.$$

From here, taking into account (2.6), we see that

$$|E \max_i X_{0i} - E \max_i X_{1i}| \leqslant \frac{\beta\delta}{4} + \frac{2\ln d}{\beta}.$$

Now setting $\beta = \sqrt{8\delta^{-1}\ln d}$, we obtain (2.4).

Remark 2.4. This proof is a modification of the proof from Chatterjee [2008].

Remark 2.5. This approach also allows one to prove the generalization of the Fernique-Sudakov inequality for a general functional $f(\max_{i,j}(x_i - x_j))$ with a positive increasing concave function f.

2.3 Berman's Inequality and its Generalizations

Now let us consider an identity similar to (2.3), but for non-smooth functions f. We have seen similar situations already with Slepian's inequality where indicator functions were used, and also when considering Fernique-Sudakov's inequality with the underlying functions differentiable almost everywhere but not twice-differentiable. In those cases we used approximations by sufficiently smooth functions, but such an approach is not always

the most convenient one. Let us now prove an analogue to (2.3) for a class
of indicator functions directly, which would be useful in studying excursions
of trajectories of Gaussian functions above a high level.

Let u be a real value, the level of a high barrier. Let us denote by $\mathcal{A}$ the
algebra of sets in $\mathbb{R}^d$ generated by the collection of sets

$$\{\mathbf{x} = (x_1,\ldots,x_d) \in \mathbb{R}^d : x_k \leqslant u\}, \{\mathbf{x} = (x_1,\ldots,x_d) \in \mathbb{R}^d : x_k < u\}, k = 1,\ldots,d.$$

Theorem 2.6. *Let* $\mathbf{X}_0 = (X_{01},\ldots,X_{0d})$ *and* $\mathbf{X}_1 = (X_{11},\ldots,X_{1d})$ *be two independent zero-mean Gaussian vectors with covariances* r_{0ij} *and* r_{1ij}, *re-
spectively,* $r_{0ii} = r_{1ii} = 1$ *for all* $i = 1,...,d$, *and* $|r_{0ij}| < 1$ *for all* $i \neq j$. *Then
for any* u *and* $A \in \mathcal{A}$,

$$P(\mathbf{X}_1 \in A) - P(\mathbf{X}_0 \in A)$$

$$= \sum_{k,l=1,\ k>l}^{d} (r_{1kl} - r_{0kl}) \int_0^1 \varphi(u,u,r_{hkl}) \sum_{\mu,\nu=0,1} (-1)^{\nu+\mu}$$

$$\times P(\mathbf{X}_h \in A \cap \pi_{h\mu k} \cap \pi_{h\nu l} \mid X_{hk} = X_{hl} = u)dh,$$

where we have denoted

$$\pi_{h\nu l} = \{(-1)^\nu (X_{hl} - u) > 0\}, \ \mathbf{X}_h = \sqrt{h}\mathbf{X}_1 + \sqrt{1 - h}\mathbf{X}_0,$$

and $\varphi(u,u,r)$ *is the bi-variate Gaussian density with zero means, unit variances and the covariance* r.

Remark 2.7. This identity also holds for dependent $\mathbf{X}_0$ and $\mathbf{X}_1$, but the
vector $\mathbf{X}_h$ should be constructed from independent copies of $\mathbf{X}_0$ and $\mathbf{X}_1$.

Proof. As we have done a few times before, let us first suppose that $\mathbf{X}_0$
and $\mathbf{X}_1$ are non-degenerate, that is their distributions have densities that
we denote by $\varphi_0(\mathbf{x})$ and $\varphi_1(\mathbf{x})$, $\mathbf{x} \in \mathbb{R}^d$. Then $\mathbf{X}_h$ has a density as well, as
the set of all positive-definite matrices is convex. The set A is a union of
d-cones,

$$\Pi_\nu = \{(-1)^{\nu_1}(x_1 - u) > 0,\ldots,(-1)^{\nu_d}(x_d - u) > 0\},$$

where $\nu = (\nu_1,\ldots,\nu_d)$, $\nu_i = 0$ or $= 1$. Some of the strict ">" inequalities
can be replaced by weak ones "$\geqslant$", but it does not matter in reality as the
densities exist, so the probabilities do not change from one case to the next.

Let us denote by $\nu(A)$ the set of indices that correspond to A so that

$$A = \cup_{\nu \in \nu(A)} \Pi_\nu.$$

Using the expression for r_{hkl} we have,

$$P(\mathbf{X}_1 \in A) - P(\mathbf{X}_1 \in A) = \int_0^1 \frac{d}{dh} P(\mathbf{X}_h \in A) dh$$

$$= \int_0^1 \frac{d}{dh} P(\mathbf{X}_h \in \cup_{\nu \in \nu(A)} \Pi_\nu) dh$$

$$= \int_0^1 \frac{d}{dh} \int_{\cup_{\nu \in \nu(A)} \Pi_\nu} \varphi_h(\mathbf{x}) d\mathbf{x}\, dh$$

$$= \int_0^1 \left(\sum_{\nu \in \nu(A)} \int_{\Pi_\nu} \frac{d}{dh} \varphi_h(\mathbf{x}) d\mathbf{x} \right) dh$$

$$= \int_0^1 \left(\sum_{\nu \in \nu(A)} \int_{\Pi_\nu} \sum_{k>l} \frac{\partial \varphi_h(\mathbf{x})}{\partial r_{hkl}} \frac{dr_{hkl}}{dh} d\mathbf{x} \right) dh$$

$$= \sum_{k>l} (r_{1kl} - r_{0kl}) \int_0^1 \sum_{\nu \in \nu(A)} \int_{\Pi_\nu} \frac{\partial \varphi_h(\mathbf{x})}{\partial r_{hkl}} d\mathbf{x}\, dh. \qquad (2.7)$$

For an arbitrary pair of positive integers k,l, let us split the set $\nu(A)$ into four subsets such that $\nu_k = 0$ or 1, and $\nu_l = 0$ or 1. Let us denote these four subsets by $\nu_{kl}(A)$. It should be noted that some of them could be empty. Integrating by parts, remembering (2.1) and using again the symbol $\frac{d\mathbf{x}}{dx_k dx_l}$ to denote the product of the differentials dx_i excluding dx_k and dx_l, we obtain that

$$\int_{\Pi_\nu} \frac{\partial \varphi_h(\mathbf{x})}{\partial r_{hkl}} d\mathbf{x} = \int_{\Pi_\nu} \frac{\partial^2 \varphi_h(\mathbf{x})}{\partial x_k \partial x_l} d\mathbf{x} = (-1)^{\nu_k+\nu_l} \int_{\Pi_\nu} \varphi_h(\mathbf{x})|_{x_k = x_l = u} \frac{d\mathbf{x}}{dx_k dx_l}$$

$$= (-1)^{\nu_k+\nu_l} P(\mathbf{X}_h \in \Pi_\nu \mid X_{hk} = X_{hl} = u) \varphi(u,u,r_{hkl}).$$

The sets $\nu_{kl}(A)$ are pairwise disjoint, and $\nu(A) = \cup_{\nu_k,\nu_l=0,1} \nu_{kl}(A)$. Therefore

$$\sum_{\nu \in \nu(A)} \int_{\Pi_\nu} \frac{\partial \varphi_h(\mathbf{x})}{\partial r_{hkl}} d\mathbf{x} = \sum_{\nu_k,\nu_l=0,1} \sum_{\nu \in \nu_{kl}(A)} \int_{\Pi_\nu} \frac{\partial \varphi_h(\mathbf{x})}{\partial r_{hkl}} d\mathbf{x}$$

$$= \varphi(u,u,r_{hkl}) \sum_{\nu_k,\nu_l=0,1} (-1)^{\nu_k+\nu_l} \sum_{\nu \in \nu_{kl}(A)} P(\mathbf{X}_h \in \Pi_\nu \mid X_{hk} = X_{hl} = u)$$

$$= \varphi(u,u,r_{hkl}) \sum_{\nu_k,\nu_l=0,1} (-1)^{\nu_k+\nu_l} P(\mathbf{X}_h \in \cup_{\nu \in \nu_{kl}(A)} \Pi_\nu \mid X_{hk} = X_{hl} = u)$$

$$= \varphi(u,u,r_{hkl}) \sum_{\nu_k,\nu_l=0,1} (-1)^{\nu_k+\nu_l} P(\mathbf{X}_h \in \cup_{\nu \in \nu_{kl}(A)} \Pi_\nu \mid X_{hk} = X_{hl} = u) \quad (2.8)$$

By the definition of $\nu_{kl}(A)$,

$$P(\mathbf{X}_h \in \cup_{\nu \in \nu_{kl}(A)} \Pi_\nu \mid X_{hk} = X_{hl} = u)$$

$$= P(\mathbf{X}_h \in A \cap \pi_{h\mu k} \cap \pi_{h\nu l} \mid X_{hk} = X_{hl} = u),$$

which, together with (2.7) and (2.8), prove the theorem under the assumption that the densities exist. To relax this assumption we use our usual trick – we introduce two n-dimensional independent standard Gaussian vectors Λ_0 and Λ_1 and let

$$\mathbf{X}_0^\varepsilon = \sqrt{1-\varepsilon}\mathbf{X}_0 + \sqrt{\varepsilon}\Lambda_0 \quad \text{and} \quad \mathbf{X}_1^\varepsilon = \sqrt{1-\varepsilon}\mathbf{X}_1 + \sqrt{\varepsilon}\Lambda_1.$$

Then we apply the result of the theorem to these two non-degenerate Gaussian vectors, and let ε go to zero. Since the absolute values of all covariances of $\mathbf{X}_0$ are strictly less then one, all two-dimensional distributions are non-degenerate for almost all h, $0 \leqslant h < 1$, and the theorem is proved in its full generality.

Corollary 2.8. *Let us assume that all the conditions of Theorem 2.6 are fulfilled, except the condition of independence of $\mathbf{X}_1$ and $\mathbf{X}_0$. Then for any u and $A \in \mathcal{A}$,*

$$|P(\mathbf{X}_1 \in A) - P(\mathbf{X}_0 \in A)|$$

$$\leqslant \frac{1}{\pi} \sum_{k,l=1,\ k>l}^{d} \frac{|r_{1kl} - r_{0kl}|}{\sqrt{1 - r_{1kl}^2 \vee r_{0kl}^2}} \exp\left(-\frac{u^2}{1 + r_{1kl} \vee r_{0kl}}\right).$$

Proof. Notice that on the right-hand side of the statement of Theorem 2.6 we have two probabilities appearing with the sign "plus", and two others appearing with the sign "minus". It follows that the sum of these probabilities is at most two. The corollary follows using the obvious inequality for the bi-variate Gaussian density.

Corollary 2.9 (Berman's comparison inequality). *Let the assumptions of Theorem 2.6 be fulfilled, and let $\mathbf{X}_0$ be a Gaussian standard vector. Then for all u,*

$$|P(\max_{i=1,\ldots,d} X_{0i} \leqslant u) - P(\max_{i=1,\ldots,d} X_{1i} \leqslant u)|$$

$$\leqslant \frac{1}{2\pi} \sum_{k,l=1,\ k>l}^{d} \frac{|r_{1kl}|}{\sqrt{1 - r_{1kl}^2}} \exp\left(-\frac{u^2}{1 + r_{1kl}}\right).$$

Proof. Notice that on the right-hand side of the statement of Theorem 2.6 there is only one non-zero probability. The result follows from elementary estimates for a bi-variate Gaussian density.

2.4 Tails of Gaussian Vectors

Proposition 2.10. *For a standard Gaussian random variable X and all $x > 0$,*

$$\frac{1}{\sqrt{2\pi}x}e^{-x^2/2}\left(1-\frac{1}{x^2}\right) \leqslant \mathrm{P}(X>x) \leqslant \frac{1}{\sqrt{2\pi}x}e^{-x^2/2} \quad \text{with } x>0.$$

Proof. Integrating by parts we obtain,

$$\int_x^\infty \frac{1}{y^2}e^{-y^2/2}dy = -\int_x^\infty e^{-y^2/2}d\frac{1}{y} = -\frac{1}{y}e^{-y^2/2}\Big|_x^\infty - \int_x^\infty e^{-y^2/2}dy$$

$$= \frac{1}{x}e^{-x^2/2} - \sqrt{2\pi}\mathrm{P}(X>x).$$

From here,

$$\sqrt{2\pi}\mathrm{P}(X>x) \leqslant \frac{1}{x}e^{-x^2/2}.$$

Further,

$$\int_x^\infty \frac{1}{y^2}e^{-y^2/2}dy \leqslant \frac{1}{x^3}\int_x^\infty ye^{-y^2/2}dy = \frac{1}{x^3}e^{-x^2/2},$$

which gives the required lower bound.

Proposition 2.11. *Let* $(X,Y)^\top$ *be a Gaussian bi-variate vector with zero mean, unit variances of its components and the covariance* r, $|r|<1$. *Then*

$$\mathrm{P}(X>x, Y>x) \sim \frac{(1+r)^{3/2}}{2\pi x^2\sqrt{1-r}}\exp\left(-\frac{x^2}{1+r}\right) \quad \text{where } x\to\infty.$$

Proof. Since the probability density $\varphi(x,y,r)$ of this vector exists, we have,

$$\frac{d}{dr}\mathrm{P}(X>x, Y>x) = \varphi(x,x.r),$$

hence,

$$\mathrm{P}(X>x, Y>x) = \int_0^r \varphi(x,x,y)dy + \mathrm{P}(X>x)^2.$$

Notice that

$$\varphi(x,x,y) = \frac{1}{2\pi\sqrt{1-y^2}}\exp\left(-\frac{x^2}{1+y}\right).$$

Now we have,

$$\frac{1}{2\pi\sqrt{1-r^2}}\exp\left(-\frac{x^2}{1+r}\right)\int_0^r \frac{\sqrt{1-r^2}}{\sqrt{1-y^2}}\exp\left(\frac{x^2}{1+r}-\frac{x^2}{1+y}\right)dy$$

$$= \frac{1}{2\pi\sqrt{1-r^2}}\exp\left(-\frac{x^2}{1+r}\right)\int_0^r \frac{\sqrt{1-r^2}}{\sqrt{1-y^2}}\exp\left(-\frac{x^2(r-y)}{(1+r)(1+y)}\right)dy$$

$$= \frac{1}{2\pi\sqrt{1-r^2}}\exp\left(-\frac{x^2}{1+r}\right)\frac{1}{x^2}$$

$$\times \int_0^{rx^2} \frac{\sqrt{1-r^2}}{\sqrt{1-(r-z/x^2)^2}}\exp\left(-\frac{z}{(1+r)(1+r-z/x^2)}\right)dz,$$

where we applied the change of variables $y = r - z/x^2$. By the dominating convergence theorem, the integral is equal to

$$\int_0^\infty \exp\left(-\frac{z}{(1+r)^2}\right) dz(1 + o(1)) = (1 + r)^2(1 + o(1)) \text{ as } x \to \infty.$$

Proposition 2.12. *Let* $\mathbf{X} = (X_1, \ldots, X_d)$ *be a Gaussian vector such that* $\mathbf{EX} = \mathbf{0}$, *all the component variances are equal to one and all the covariances are less than one,* $|r_{ij}| < 1$. *Then*

$$P(\max_i X_i > x) \sim dP(X_1 > x) \text{ as } x \to \infty.$$

Proof. We have,
$$P(\max_i X_i > x) \leqslant dP(X_1 > x)$$

and
$$P(\max_i X_i > x) \geqslant dP(X_1 > x) - \sum_{i \neq j} P(X_i > x, X_j > x).$$

Applying Proposition 2.11 to the probabilities under the sum and Proposition 2.10 to the single probability on the right, we see that the sum is infinitely smaller than the first probability on the right.

3

Ergodicity of Stationary Sequences

This lecture consists of two parts. The first covers some essential facts, mostly without proofs, on mixing properties of Gaussian stationary sequences. The second part looks into mixing properties of the *trace* (the process of signs) of a Gaussian stationary sequence. Moreover, we establish a Poisson limit theorem for large values of Gaussian stationary sequence.

Let $X(k) = X(k,\omega)$, $k \in \mathbb{Z}$, $\omega \in \Omega$, be a Gaussian stationary sequence, sometimes called a time series. By stationarity here we mean that its expected value $EX(k) = a$ is constant, and its covariance function $r(k,l) = E(X(k) - a)(X(l) - a)$ depends only on the difference of the arguments, $r(k,l) = r(l - k)$. For general, non-Gaussian, random processes such a notion of stationarity is called *weak stationarity*, in contrast to strong stationarity where all finite-dimensional distributions do not depend on time shifts. In the Gaussian case both notions coincide as finite-dimensional distributions depend only on means and covariances.

By the Bochner-Khinchin theorem, the following spectral representation can be written down,

$$r(k) = \int_{-\pi}^{+\pi} e^{i\lambda k} dF(\lambda),$$

where $F(\lambda)$ is an increasing, bounded, right-continuous function with $F(-\pi) = 0$. This function is called the *spectral function* of $X(k)$, and the corresponding measure $F(d\lambda) = dF(\lambda)$ is called its *spectral measure*. The names come from a spectral representation of the process $X(k)$ as a stochastic integral,

$$X(k) = \int_{-\pi}^{+\pi} e^{i\lambda k} \zeta(d\lambda),$$

where $\zeta(\lambda)$ is a zero mean stochastic process with orthogonal increments such that

$$E\zeta(d\lambda)\overline{\zeta(d\mu)} = \begin{cases} dF(\lambda), & \text{if } \lambda = \mu, \\ 0, & \text{if } \lambda \neq \mu. \end{cases}$$

The Bochner-Khinchin theorem also states that the spectral function is unique. It is worth noting that the above representation holds for any, and not just Gaussian, stationary sequences.

Let $\mathcal{A}$ be a σ-algebra generated by all the values of $X(k)$ for $k \in \mathbb{Z}$, and $H = H(X)$ be a space of all random $\mathcal{A}$-measurable variables with finite variances. Let us introduce the shift operator $T_k, k \in \mathbb{Z}$, in H by the relation $T_k X(l) = X(l + k), l \in \mathbb{Z}$. A random variable $X \in H$ is called *invariant with respect to shifts*, if $T_k X = X$ for all k with probability one. A constant is obviously invariant. If all invariant random variables in H are constant with probability one, then the random sequence $X(k)$ is called *ergodic*. For the ergodic random sequences, and only for them, for any $\xi \in H$, we have that

$$\frac{1}{T} \sum_{k=0}^{T} T_k \xi \to \mathrm{E}\xi$$

in probability as $T \to \infty$. The following theorem holds.

Theorem 3.1 (G. Maruyama). *A Gaussian stationary sequence $X(k)$, $k \in \mathbb{Z}$, is ergodic if and only if F is continuous.*

One says that $X(k)$ satisfies the *mixing property* if, for any $X, Y \in H$, $\mathrm{E}XT_kY \to \mathrm{E}X\mathrm{E}Y$ as $k \to \infty$.

Theorem 3.2 (K. Itô). *A Gaussian stationary sequence $X(k), k \in \mathbb{Z}$, satisfies the mixing property if and only if $r(k) \to 0$ as $k \to \infty$.*

The above two theorems can be proved by expanding random variables from H into multiple Itô's stochastic integrals. Outlines of the proofs can be found in K. Itô's book Ito et al. [2004], volume 1.

Let $\mathcal{A}(X(\cdot); S, T)$ be a σ-algebra generated by random variables $X(k)$, $S < k < T$. A strongly stationary random sequence $X(k)$ is said to have the *strong mixing property* (or Rosenblatt's mixing, or α-mixing), if

$$\alpha(\tau) = \sup_{A \in \mathcal{A}(X(\cdot); -\infty, 0), B \in \mathcal{A}(X(\cdot); \tau, +\infty)} (\mathrm{P}(AB) - \mathrm{P}(A)\mathrm{P}(B)) \to 0$$

as $\tau \to \infty$.

Theorem 3.3. *If the spectral measure $F(d\lambda)$ of a Gaussian stationary sequence $X(k), k \in \mathbb{Z}$, has a density $f(\lambda), \lambda \in [-\pi, \pi]$, with respect to the Lebesque measure, and $\inf_{\lambda \in [-\pi, \pi]} f(\lambda) > 0$ for at least one version of f, then this sequence is strongly mixing. If $f(\lambda) \equiv 0$ in a non-empty open set, then this sequence is not strongly mixing.*

Let us now present one of the simpler versions of the central limit theorem under the strong mixing condition. It will be useful later on.

Theorem 3.4. *Let a stationary sequence $X(k)$ satisfy the strong mixing condition, and $\sum_{n=1}^{\infty} \alpha(n) < \infty$. Let the random variables $X(k)$ be bounded with probability one, $|X(k)| < c_0 < \infty$. Then*

$$\sigma^2 := EX^2(0) + 2\sum_{j=1}^{\infty} EX(0)X(j) < \infty.$$

If, furthermore, $\sigma \neq 0$, then

$$P\left(\frac{1}{\sigma\sqrt{n}}\sum_{j=1}^{n} X(j) < z\right) \to \Phi(z)$$

as $n \to \infty$.

The proofs of these and many related results can be found in Rozanov [1967] and Ibragimov and Rozanov [1978].

Having reviewed relevant mixing results, let us now move on to the the second part of this lecture where we want to apply Theorem 2.6 to mixing conditions for sign processes generated by a Gaussian stationary sequence. Let $X(k), k \in \mathbb{Z}$, be a Gaussian stationary sequence. Let us consider its trace at level u,

$$U(k) = \mathbf{I}_{X(k)>u}, \ k \in \mathbb{Z},$$

where $\mathbf{I}_{(\cdot)}$ is an indicator function.

Theorem 3.5. *Let $r(k)$ be the covariance function of a Gaussian stationary sequence $X(k), k \in \mathbb{Z}$. If*

$$\sum_{k=1}^{\infty} |kr(k)| < \infty,$$

then for any level u, the random sequence $U(k)$ is strongly mixing and, moreover, the strong mixing coefficient α satisfies

$$\alpha(\tau) \leqslant \frac{2}{\pi} \frac{1}{\sqrt{r(0) - \sup_{k\geqslant\tau} r^2(k)}} \exp\left(-\frac{(u-m)^2}{r(0) + \sup_{k\geqslant\tau} r(k)}\right) \sum_{k=\tau}^{\infty} |kr(k)|,$$

where m is the mean of the sequence.

Proof. Without loss of generality we may assume that $EX(k) = 0$ and $EX^2(k) = 1$. Let $T > 0$ and $\tau > 0$ be arbitrary positive integers, and let us consider two random events $A \in \mathcal{A}(U(\cdot); -T, 0)$ and $B \in \mathcal{A}(U(\cdot); \tau, \tau + T)$. Let us furthermore introduce two independent copies $X_0(k)$ and $X_1(k)$ of the sequence $X(k)$, and denote by P_0 the Gaussian measure generated by the $(2T + 2)$-vector

$$(X_0(k), -T \leqslant k \leqslant 0, X_1(l), \tau \leqslant l \leqslant \tau + T).$$

Then $P_0(AB) = P(A)P(B)$. By Corollary 2.8,

$$|P(AB) - P(A)P(B)| = |P(AB) - P_0(AB)|$$

$$\leqslant \frac{2}{\pi} \sum_{l\in[-S,0],k\in[\tau,\tau+T]} \frac{r(k-l) - r_0(k-l)}{\sqrt{1-r^2(k-l)}} \exp\left(-\frac{u^2}{1+r(k-l)}\right)$$

$$\leqslant \frac{2}{\pi} \frac{1}{\sqrt{1-\sup_{l\geqslant\tau} r^2(l)}} \exp\left(-\frac{u^2}{1+\sup_{l\geqslant\tau} r(l)}\right) \sum_{k=\tau}^{\infty} |kr(k)|.$$

Since this bound is uniform in time T and the events A and B, the theorem follows.

We can apply Theorem 3.4 to establish the asymptotic normality of the number of level exceedances by a Gaussian stationary sequence.

Theorem 3.6. *Let the covariance function $r(k)$ of a Gaussian stationary sequence $X(k)$, $k \in \mathbb{Z}$, be such that*

$$\sum_{k=1}^{\infty} |k^2 r(k)| < \infty.$$

In addition let us assume that the variance of the number of exceedings tends to infinity,

$$\mathrm{Var} \sum_{j=1}^{T} U(j) \to \infty$$

as $T \to \infty$. Then, as $T \to \infty$, we have,

$$\mathrm{Var} \sum_{j=1}^{T} U(j) \sim \sigma^2 T, \ \sigma > 0,$$

and for any z,

$$P\left(\frac{1}{\sigma\sqrt{T}} \sum_{j=1}^{T} (U(k) - P(X(1) > u)) < z\right) \to \Phi(z).$$

Proof. The first assertion follows from Leonov's theorem that states that for any stationary random sequence $\xi(k)$ with a correlation function $R(k)$, the inequality

$$\sum_{k=1}^{\infty} k|R(k)| < \infty$$

implies

$$\mathrm{Var} \sum_{k=1}^{T} \xi(k) = T \sum_{k=-\infty}^{\infty} R(k) - 2 \sum_{k=1}^{\infty} kR(k) + o(1)$$

as $T \to \infty$. The proof of this theorem is straightforward, and a good exercise for the reader, see also Leonov [1961]. Since by the conditions, the variance of the sum of $U(k)$'s tends to infinity and, by the comparison theorem, the covariance function of the sequence $U(k)$ is dominated by a constant times $|r(k)|$, Leonov's theorem ensured that the variance of the sum increases proportionally to T. Furthermore, by Theorem 3.5, Theorem 3.4 conditions are fulfilled for $U(k)$, and the result follows.

Now let us study the behavior of $U(k)$ when u is large. To fix ideas, let us suppose for a second that all $X(k)$ are independent and, of course, identically distributed, so that $r(k) = 0$ for all $k > 0$. For simplicity we again assume that $EX(k) = 0$ and $EX^2(k) = 1$. Hence $U(k)$ is a Bernoulli sequence with the parameter $P(U(k) = 1) = 1 - \Phi(u)$, where $\Phi(x)$ is the standard Gaussian distribution function as before. By the classic Poisson limit theorem, if $n \to \infty$ and $u \to \infty$ in such a way that $n(1 - \Phi(u)) \to \lambda > 0$, then the sequence

$$\eta_X(u, n) = \sum_{k=1}^{n} U(k)$$

tends in distribution to a Poisson random variable with the parameter λ. It turns out that the same happens in the case of dependent $X(k)$.

Theorem 3.7. *Let the covariance function $r(k)$ of a Gaussian stationary sequence $X(k)$, $EX(k) = 0$, $EX^2(k) = 1$, satisfy the Berman's condition,*

$$r(k) \ln k \to 0 \ as \ k \to \infty. \tag{3.1}$$

Let $n \to \infty$ and $u \to \infty$ in such a way that $(1 - \Phi(u))n \to \lambda > 0$. Then the sequence $\eta_X(u, n)$ converges in distribution to a Poisson random variable with the parameter λ.

Proof. Let $X_0(k)$ be a sequence of independent Gaussian standard variables. Notice that $\sup_{k \geq 1} |r(k)| =: \rho(1) < 1$. Otherwise $r(k)$ would equal to one for infinitely many k, which contradicts the assumed relation[1] $r(k) \to 0$ as $k \to \infty$. Hence we can apply the comparison theorem to the vectors $(X(k), k = 1, \ldots, n)$ and $(X_0(k), k = 1, \ldots, n)$. For any integer $m > 0$, we have,

$$|P(\eta_X(u, n) = m) - P(\eta_{X_0}(u, n = m)|$$

$$\leqslant 2 \sum_{i,j=1, i>j}^{n} |r(i - j)| \int_0^1 \frac{1}{2\pi\sqrt{1 - h^2 r^2(i-j)}} \exp\left(-\frac{u^2}{1 + hr(i-j)}\right) dh$$

$$\leqslant \frac{1}{\pi} \sum_{k=1}^{n} k|r(k)| \int_0^1 \frac{1}{\sqrt{1 - h^2 r^2(k)}} \exp\left(-\frac{u^2}{1 + hr(k)}\right) dh. \tag{3.2}$$

Let us show that the right hand part in (3.2) tends to zero. By conditions of the theorem and the asymptotic behavior of $1 - \Phi(u)$,

---

[1] The reader should prove this!

$$\lim_{u \to \infty} \frac{n}{\sqrt{2\pi u}} \exp(-u^2/2) = \lambda.$$

Taking logarithm we get,

$$\lim_{u \to \infty} (\ln n - u^2/2 - \ln u) = \ln \sqrt{2\pi}\lambda. \tag{3.3}$$

From here, $u^{-1}\sqrt{2\ln n} \to 1$ as $u \to \infty$ (divide both parts by $u^2/2$). Moreover, from the conditions of the theorem it follows that

$$\lim_{k \to \infty} \rho(k) \ln k = 0, \text{ where } \rho(k) = \sup_{t \geq k} r(t).$$

We split the sum on the right-hand side of (3.2) into two parts, for $k = 1$ to $[\sqrt{n}]$, and from $[\sqrt{n}] + 1$ to n. We have,

$$\sum_{k=1}^{[\sqrt{n}]} k|r(k)| \int_0^1 \frac{1}{\sqrt{1 - h^2 r^2(k)}} \exp\left(-\frac{u^2}{1 + hr(k)}\right) dh$$

$$\leq \sqrt{n}\sqrt{n}\rho(1)\frac{1}{\sqrt{1 - \rho^2(1)}} \exp\left(-\frac{u^2}{1 + \rho(1)}\right)$$

$$= \frac{n}{u}e^{-u^2/2}O\left(u \exp\left(-\frac{u^2(1 - \rho(1))}{2(1 + \rho(1))}\right)\right) \to 0$$

as $u \to \infty$. For the second sum,

$$\sum_{k=[\sqrt{n}]+1}^{n} k|r(k)| \int_0^1 \frac{1}{\sqrt{1 - h^2 r^2(k)}} \exp\left(-\frac{u^2}{1 + hr(k)}\right) dh$$

$$\leq n^2 |\rho([\sqrt{n}])|\frac{1}{\sqrt{1 - \rho^2([\sqrt{n}])}} \exp\left(-\frac{u^2}{1 + \rho([\sqrt{n}])}\right)$$

$$= \frac{1}{\sqrt{1 - \rho^2([\sqrt{n}])}} \exp(2\ln n - u^2 - 2\ln u) \exp\left(\frac{\rho([\sqrt{n}])u^2}{1 + \rho([\sqrt{n}])}\right) u^2 |\rho([\sqrt{n}])|.$$

Since $u^{-1}\sqrt{2\ln n} \to 1$ as $n \to \infty$, $u^2 |\rho([\sqrt{n}])| \to 0$ and, thanks to (3.3), the whole of the product tends to zero. As $\eta_{X_0}(u,n)$ converges to a Poisson random variable with parameter λ, the theorem follows.

To round off this lecture, let us establish a limit theorem for the maximum of a Gaussian stationary sequence over an unboundedly increasing time interval. Letting $u = u_n = \sqrt{2\ln n} + \alpha_n/\sqrt{2\ln n}$ and putting this in (3.3), it is easy to check that in order to get a non-degenerate limit one can take

$$u_n = \sqrt{2\ln n} - \frac{\frac{1}{2}\ln\ln n + \ln\sqrt{\frac{\pi}{2}\lambda}}{\sqrt{2\ln n}}.$$

Corollary 3.8 (Berman's Theorem). *Under Berman's condition (3.1), we have that for*

$$a_n = \sqrt{2\ln n}, \ b_n = \sqrt{2\ln n} - \ln\sqrt{4\pi\ln n}/\sqrt{2\ln n},$$

and all x,

$$\lim_{n\to\infty} \mathrm{P}\left(a_n\left(\max_{1\leqslant t\leqslant n} X(t) - b_n\right) \leqslant x\right) = \exp\left(-e^{-x}\right).$$

Let us note that the normalizing sequences a_n, b_n are the same as in the case of independent standard Gaussian variables, that is, under the conditions of the Fisher-Gnedenko-Tippet theorem. In order to prove the corollary, it is sufficient to see that the events $\{\eta_X(u,n) = 0\}$ and $\{\max_{1\leqslant k\leqslant n} X(k) < u_n\}$ coincide.

4

Zero-One Law

In this lecture we consider Gaussian random variables taking values in possibly infinite-dimensional spaces and, in particular, in functional spaces. To do that effectively we need to revisit definitions from Lecture 1 with a more general viewpoint.

Let T be an arbitrary set and $(\Omega, \mathcal{F}, P)$ a probability space. A family of real-valued random variables $\mathbf{X} = (X(t,\omega), t \in T, \omega \in \Omega)$ is called Gaussian if for any finite subset $T_0 \subset T$, the random vector $\mathbf{X}_0 = (X(t,\omega), t \in T_0, \omega \in \Omega)$ is Gaussian. In this lecture we assume that the expected value of $X(t,\omega)$ is zero for all t. Then the covariance function $r(s,t) = EX(s)X(t)$, $(s,t) \in T \times T$, alone determines all finite dimensional distributions. So far we have considered Gaussian random vectors in finite-dimensional Euclidean spaces. Now we would like to study Gaussian vectors taking values in arbitrary vector spaces and, in particular, in arbitrary functional spaces. From this point of view, a Gaussian function, which is nothing but a Gaussian vector in the space of functions over a subset T of the real line, is the subject of our attention in this lecture, as well as others to follow.

4.1 Alternative Definition of a Gaussian Vector

Let us remind the reader some standard definitions from the axiomatic probability theory. A couple $(\mathbb{A}, \mathcal{A})$, with $\mathbb{A}$ a set and $\mathcal{A}$ a σ-algebra of its subsets, is called a measurable space. A simple example is given by $(\mathbb{R}^d, \mathcal{B}^d)$, the Euclidean space and the σ-algebra of its Borel subsets.

Definition 4.1. *Let $\mathbb{E}$ be a vector space over the field of real numbers, and $\mathcal{E}$ be a σ-algebra of its subsets. We say that $(\mathbb{E}, \mathcal{E})$ is a* measurable vector space *if the operations of multiplication by a scalar and addition of two vectors are measurable mappings of the spaces $(\mathbb{R} \times \mathbb{E}, \mathcal{B} \times \mathcal{E})$ and $(\mathbb{E} \times \mathbb{E}, \mathcal{E} \times \mathcal{E})$, respectively, into the space $(\mathbb{E}, \mathcal{E})$.*

Let us remind the reader that a mapping ξ of a measurable space $(\mathbb{A}, \mathcal{A})$ into a measurable space $(\mathbb{X}, \mathcal{X})$ is called measurable if, for any $X \in \mathcal{X}$, $\xi^{-1}(X) \in \mathcal{A}$. If $(\mathbb{A}, \mathcal{A}) = (\Omega, \mathcal{F})$, we just call it a random variable taking values in $\mathbb{X}$, or simply a random variable in $\mathbb{X}$. A measure $\mu(X) := P(\xi^{-1}(X))$ on $\mathcal{X}$ is called the distribution of ξ. A random variable η is called a copy of ξ if both of them have the same distribution. Random variables ξ and η are called independent if the events $\xi^{-1}(X)$ and $\eta^{-1}(Y)$ are independent for any $X, Y \in \mathcal{X}$, that is, $\mu(X \cap Y) = \mu(X)\mu(Y)$. Let us state a simple proposition.

Proposition 4.2. *A space $(\mathbb{E}, \mathcal{E})$, where $\mathbb{E}$ is a vector space over the field of real numbers, and $\mathcal{E}$ is a σ-algebra of its subsets, is a measurable vector space if and only if for any measurable space $(\mathbb{A}, \mathcal{A})$, and a pair (X, Y) of measurable mappings from $(\mathbb{A}, \mathcal{A})$ to $(\mathbb{E}, \mathcal{E})$, and any pair (λ, μ) of measurable mappings from $(\mathbb{A}, \mathcal{A})$ to $(\mathbb{R}, \mathcal{B})$, the mapping $\lambda X + \mu Y$ from $(\mathbb{A}, \mathcal{A})$ to $(\mathbb{E}, \mathcal{E})$ is measurable.*

Proof. The "if" part follows from the definition directly. To prove the "only if" part, we set the space $(\mathbb{A}, \mathcal{A})$ to be, first, $(\mathbb{R} \times \mathbb{E}, \mathcal{B} \times \mathcal{E})$, and then, next, $(\mathbb{E} \times \mathbb{E}, \mathcal{E} \times \mathcal{E})$.

Definition 4.3. *Let $(\Omega, \mathcal{F}, P)$ be a probability space, $(\mathbb{E}, \mathcal{E})$ be a measurable vector space. A measurable mapping X from $(\Omega, \mathcal{F})$ to $(\mathbb{E}, \mathcal{E})$ is a Gaussian random vector in $\mathbb{E}$ if for any pair (X_1, X_2) of independent copies of X, and any pair (s, t) of real numbers such that $s^2 + t^2 = 1$, the pair $(sX_1 + tX_2, tX_1 - sX_2)$ is also a pair of independent copies of X.*

One should be careful with the definition of a Gaussian vector given above, as the following, rather odd, example demonstrates. The main lesson here is that the σ-algebra $\mathcal{E}$ needs to be sufficiently rich for the definition to be meaningful.

Example. Let $\mathbb{H}$ be a proper nonzero subspace of $\mathbb{E}$, and $\mathcal{E}$ be the σ-algebra generated by sets $\{e + \mathbb{H}\}$, $e \in \mathbb{E}$. Then any random vector X from $(\mathbb{E}, \mathcal{E})$, taking values in $\mathbb{H}$, is Gaussian. Indeed, for any $B \in \mathcal{E}$, the probability of the event $X^{-1}(B)$ equals 0 or 1, that is, any two copies of X_1 and X_2 of X are independent, but $sX_1 + tX_2$ are also copies of X, by linearity.

We shall keep in mind this exotic example, but it is more interesting to see how the definition works on measurable spaces with a richer structure, such as Euclidean spaces.

Gaussian random variable. Let us now consider the case of $(\mathbb{E}, \mathcal{E}) = (\mathbb{R}, \mathcal{B})$. Let us find the characteristic function φ of a Gaussian random variable X in the sense of the previous definition. By this definition,

$$\mathbb{E}e^{it_1 X_1 + it_2 X_2} = \varphi(t_1)\varphi(t_2) = \mathbb{E}e^{it_1(sX_1 + tX_2) + it_2(tX_1 - sX_2)}$$

$$= \varphi(st_1 + tt_2)\varphi(tt_1 - st_2).$$

Let us now proceed to solve this functional equation. Setting $t_1 = -t_2 = u$, $s = t = 2^{-1/2}$, we get the equation $\varphi(u)\varphi(-u) = |\varphi(u)|^2 = \varphi(0)\varphi(\sqrt{2}u)$, that is, $\varphi(u) = |\varphi(u/\sqrt{2})|^2$. It follows then that φ is a real symmetric function, and $\varphi(u) > 0$ for all u. (If this were not so, there would exist a sequence $t_n \to 0$ such that $\varphi(t_n) = 0$, $n = 1,2,...$, which would put the continuity of a characteristic function at zero at odds with the fact that $\varphi(0) = 1$. Furthermore, setting $t_2 = 0$, $t_1 = u$, in the initial functional equation we have,

$$\varphi(u) = \varphi(su)\varphi(tu).$$

Let us denote $\ln \varphi(u) = \psi(u^2)$; for any $y \geqslant 0$, $\psi(y) = \psi(s^2 y) + \psi((1-s^2)y)$ or $\psi(y + z) = \psi(y) + \psi(z)$ for all $y, z \geqslant 0$. Since ψ is continuous, it is linear and $\psi(0) = 0$, that is, $\psi(x) = ax$ with $a \leqslant 0$, because of $\varphi(u) \leqslant 1$. Hence, $\varphi(x) = \exp(-\frac{1}{2}\sigma^2 x^2)$, where $\sigma^2 = -2a = \mathrm{E}X^2$.

From Definition 4.3 of a Gaussian $(\mathbb{E}, \mathcal{E})$-valued vector and the calculations above, we obtain the following result.

Proposition 4.4. *If X is a Gaussian vector with values in $(\mathbb{E}, \mathcal{E})$ and $l(e) = (l, e)$ is a linear measurable function on $(\mathbb{E}, \mathcal{E})$, then $l(X)$ is a Gaussian random variable.*

Indeed, if (X_1, X_2) is a pair of independent copies of X, then (l, X_1), (l, X_2) is a pair of independent copies of (l, X), and so the characteristic function of (l, X) satisfies the same functional equation.

Gaussian finite dimensional vector. Let us now choose $(\mathbb{E}, \mathcal{E}) = (\mathbb{R}^d, \mathcal{B}^d)$. For any $\mathbf{t} \in \mathbb{R}^d$, the random variable $(\mathbf{t}, \mathbf{X})$ is Gaussian, that is, its characteristic function is given by

$$\varphi(\mathbf{t}) = \exp\left(-\frac{1}{2}\mathrm{E}(\mathbf{t}, \mathbf{X})^2\right) = \exp\left(-\frac{1}{2}\sum_{i,j}\mathrm{E}X_i X_j t_i t_j\right) = \exp\left(-\frac{1}{2}(R\mathbf{t}, \mathbf{t})\right),$$

where $R = ||r_{ij}||_{i,j=1,...,d}$.

Let us now formulate a simple but important proposition, where we use the definition of a Gaussian vector from this lecture, i.e. Definition 4.3.

Proposition 4.5. *A finite-dimensional random vector is Gaussian if and only if any linear form applied to it is a Gaussian random variable.*

4.2 Zero-One Law for Gaussian Vectors

Theorem 4.6. *Let $(\mathbb{E}, \mathcal{E})$ be a measurable vector space and $\mathbb{F}$ be its subspace such that $\mathbb{F} \in \mathcal{E}$. Then for any Gaussian $(\mathbb{E}, \mathcal{E})$-valued vector X, one of the two statements must hold,*

$$P(X \in \mathbb{F} = 0) \quad or \quad P(X \in \mathbb{F} = 1).$$

Proof. Let (X_1, X_2) be a pair of independent copies of X. Consider the events,

$$A(\theta) = \{\omega : X_1(\omega)\cos\theta + X_2(\omega)\sin\theta \in F,$$
$$X_1(\omega)\sin\theta - X_2(\omega)\cos\theta \notin F\}, \quad \theta \in [0, \pi/2].$$

We show that if $\theta_1 \neq \theta_2$ then the events $A(\theta_1)$ and $A(\theta_2)$ are mutually exclusive. Indeed, if for some ω, $X_1(\omega)\cos\theta_1 + X_2(\omega)\sin\theta_1 \in F$ and $X_1(\omega)\cos\theta_2 + X_2(\omega)\sin\theta_2 \in F$, then, since the matrix

$$\begin{pmatrix} \cos\theta_1 & \sin\theta_1 \\ \cos\theta_2 & \sin\theta_2 \end{pmatrix}$$

is non-degenerate, $X_1(\omega) \in F$, $X_2(\omega) \in F$, and so therefore $X_1(\omega)\sin\theta - X_2(\omega)\cos\theta \in F$, that is, the events have no intersection, i.e. mutually exclusive.

Let us now continue the proof. By definition, probabilities of all $A(\theta)$ are equal to one another, and since they are mutually exclusive and there are infinitely many of them, then $P(A(\theta)) = 0$ for all θ. This includes the case of $\theta = \pi/2$, and so,

$$0 = P(A(\theta)) = P(X \in F)P(X \notin F),$$

which proves the theorem.

Let us denote by $\overline{\mathbb{R}}$ the extended real line that has been augmented with the ∞ point. Let us furthermore denote by $\overline{\mathcal{B}}$ the correspondingly extended σ-algebra.

Definition 4.7. *A measurable functional N from $(\mathbb{E}, \mathcal{E})$ to $(\overline{\mathbb{R}}, \overline{\mathcal{B}})$, is called a pseudo-seminorm, if the set $N^{-1}(\mathbb{R})$ is a linear subspace of $\mathbb{E}$, in which N induced a seminorm.*

Let us remind the reader that a seminorm is a functional n such that $n(x+y) \leq n(x) + n(y)$ and $n(ax) = |a|n(x)$ for any x, y and real a.

Corollary 4.8. *Let N be a pseudo-seminorm on $\mathbb{E}$, and let X be a Gaussian vector in $(\mathbb{E}, \mathcal{E})$. Then the following two statements hold true,*

$$P(N(X) < \infty) = 0 \quad or \quad P(N(X) < \infty) = 1;$$
$$P(N(X) = 0) = 0 \quad or \quad P(N(X) = 0) = 1.$$

Indeed, the sets $\{N(e) < \infty, e \in \mathbb{E}\}$ and $\{N(e) = 0, e \in \mathbb{E}\}$ are, by definition, linear subspaces of $\mathbb{E}$. They are also elements of $\mathcal{E}$.

Let T be a set. Let us denote by $\mathcal{B}_{\mathbb{R}^T}$ the σ-algebra of cylindrical subsets of the space $\mathbb{R}^T$ of all real functions on T, that is $\mathcal{B}_{\mathbb{R}^T}$ is a σ-algebra generated by sets $\{x(t) \in B\}$, $t \in T$, $B \in \mathcal{B}$. The space $(\mathbb{R}^T, \mathcal{B}_{\mathbb{R}^T})$ is a measurable space. As defined earlier, a Gaussian random function on T is a Gaussian random vector taking values in this space. The reader should review relevant definitions and statements at this point.

Definition 4.9. *Let* T *be a set, and* $X = \{X(t,\omega),\ t \in T\}$, *be a Gaussian random function on* T. *We say that the trajectories of this function are almost sure bounded (or, simply, X is a.s. bounded), if the set* $\{\omega : \sup_T X(t,\omega) < \infty\}$ *is an event (that is, it belongs to* $\mathcal{F}$ *), and its probability is equal to one.*

Let $S \subset T$ be a topological space and $C(S)$ be a space of all continuous functions on S.

Definition 4.10. *The trajectories of X are almost sure continuous on S (in short, X is a. s. continuous on S), if the set* $\{\omega : X(\cdot,\omega) \in C(S)\}$ *is an event, and its probability is equal to one.*

Since both $C(S)$ and the set of bounded functions are subspaces of $\mathbb{R}^T$, we have the following result.

Corollary 4.11. *Any Gaussian process (Gaussian function) defined on an arbitrary topological space is either a.s. continuous or a.s. discontinuous. Any Gaussian process is either a.s. bounded or a.s. unbounded.*

Let us now discuss the connection between the usual notion of boundedness: $\sup |x(t)| < \infty$, and the property $\sup x(t) < \infty$. Let a set $S \subset T$ be countable, then it is easy to prove that the functional $N(x(\cdot)) = \sup_{t \in S} |x(t)|$ is a pseudo-seminorm in $(\mathbb{R}^T, \mathcal{B}_{\mathbb{R}^T})$. Moreover, from the definition of a Gaussian random vector it follows that the probabilities of the events

$$\{\sup_{t \in S}(X(t,\omega) < \infty\} \quad \text{and} \quad \{\sup_{t \in S}(-X(t,\omega) < \infty\}$$

coincide (since the processes have the same finite dimensional distributions). Therefore X is a.s. bounded from above on S if and only if $P(N(X) < \infty) = 1$. Moreover, a Gaussian process is a.s. bounded from above on S, if it is bounded from above with any positive probability. In other words, $P(\sup_{t \in S}(X(t,\omega) < \infty) = 0$ or 1.

5

Integrability

Properties of Gaussian distributions that we discuss in this lecture are best demonstrated with a few simple examples.

A Gaussian variable X with variance σ^2 is *exponentially integrable*, meaning that for any $\alpha < 1/2\sigma^2$,

$$\operatorname{E}\exp(\alpha X^2) < \infty.$$

Indeed, denoting $m = \operatorname{E}X$, we have,

$$\begin{aligned}
\operatorname{E}\exp(\alpha X^2) &= \frac{1}{\sqrt{2\pi}\sigma} \int \exp\left(-\frac{1}{2\sigma^2}(x-m)^2 + \alpha x^2\right) dx \\
&= \frac{1}{\sqrt{2\pi}\sigma} \int \exp\left(-\left(\frac{1}{2\sigma^2} - \alpha\right)x^2 + \frac{mx - m^2/2}{\sigma^2}\right) dx < \infty.
\end{aligned}$$

On the other hand if $\alpha \geqslant 1/2\sigma^2$, then, clearly, $\operatorname{E}\exp(\alpha X^2) = \infty$.

Let us look at another example. Let $\mathbf{X}$ be a Gaussian d-dimensional vector, with zero mean for simplicity. Denote $\sigma^2 = \max_{1\leqslant j\leqslant d} \operatorname{E}X_j^2$. We have,

$$\operatorname{E}\exp(\alpha \max_{1\leqslant j\leqslant d} X_j^2) \leqslant \sum_{j=1}^d \operatorname{E}\exp(\alpha X_j^2) < \infty$$

for $\alpha < 1/2\sigma^2$. On the other hand, for $\alpha \geqslant 1/2\sigma^2$,

$$\operatorname{E}\exp(\alpha \max_{1\leqslant j\leqslant d} X_j^2) \geqslant \max_{1\leqslant j\leqslant d} \operatorname{E}\exp(\alpha X_j^2) = \infty.$$

It turns out that a property similar to the zero-one law holds in the case of infinite dimension.

Theorem 5.1. *Let $(\mathbb{E}, \mathcal{E})$ be a measurable vector space equipped with a pseudo-seminorm N, and let X be a Gaussian vector in $(\mathbb{E}, \mathcal{E})$. Then, if $\{N(X) < \infty\}$ with non-zero probability, then there exists $\varepsilon > 0$ such that for any $\alpha < \varepsilon$,*

$$\operatorname{E}\exp(\alpha N^2(X)) < \infty.$$

Proof. We prove the theorem in the case of a centered X, so that $EX = 0$, with the general case trivially following. Let (X_1, X_2) be a pair of independent copies of X. Then

$$(N(X_1), N(X_2)) \text{ and } \left(N\left(\frac{X_1 + X_2}{\sqrt{2}}\right), N\left(\frac{X_1 - X_2}{\sqrt{2}}\right)\right)$$

are two pairs of independent copies of $N(X)$. Hence, for any x, y,

$$P(N(X) \leqslant x) P(N(X) > y) \tag{5.1}$$
$$= P\left(N\left(\frac{X_1 - X_2}{\sqrt{2}}\right) \leqslant x\right) P\left(N\left(\frac{X_1 + X_2}{\sqrt{2}}\right) > y\right).$$

By the triangle inequality,

$$N\left(\frac{X_1 + X_2}{\sqrt{2}}\right) - N\left(\frac{X_1 - X_2}{\sqrt{2}}\right) \leqslant \sqrt{2}\min(N(X_1), N(X_2)),$$

therefore

$$\left\{N\left(\frac{X_1 - X_2}{\sqrt{2}}\right) \leqslant x, N\left(\frac{X_1 + X_2}{\sqrt{2}}\right) > y\right\}$$
$$\subset \left\{N(X_1) \geqslant \frac{y - x}{\sqrt{2}}, N(X_2) \geqslant \frac{y - x}{\sqrt{2}}\right\}$$

which, together with the inequality (5.1), gives

$$P(N(X) \leqslant x) P(N(X) > y) \leqslant P\left(N(X) \geqslant \frac{y - x}{\sqrt{2}}\right)^2. \tag{5.2}$$

This estimate, in fact, gives the required exponential-quadratic inequality for the distribution tail of $N(X)$. From the zero-one law and the conditions of the theorem we obtain that $P(N(X) < \infty) = 1$. Hence we can choose y_0 so that $q := P(N(X) \leqslant y_0) > 1/2$. Let us take

$$y_{n+1} = y_0 + \sqrt{2}y_n, \ n = 0, 1, 2, \ldots \text{ and } t_n = q^{-1}P(N(X) > y_n).$$

Setting $x = y_0$, $y = y_n$ in (5.2) we get,

$$qqt_n \leqslant q^2 t_{n-1}^2,$$

that is, $t_n \leqslant t_{n-1}^2$. Therefore,

$$P(N(X) > y_n) = qt_n \leqslant qt_{n-1}^2 \leqslant \cdots \leqslant q\left(\frac{1 - q}{q}\right)^{2^n}.$$

Since the sequence $y_n = \frac{(\sqrt{2})^{n+1} - 1}{\sqrt{2} - 1} y_0$ tends to infinity as $n \to \infty$,

$$\mathrm{E}\exp(\alpha N^2(X)) \leqslant e^{\alpha y_0^2}\mathrm{P}(N(X) > y_0) + \sum_{n=1}^{\infty} e^{\alpha y_n^2}\mathrm{P}(N(X) > y_n)$$

$$\leqslant q\left[e^{\alpha y_0^2} + \sum_{n=1}^{\infty}\left(\frac{1-q}{q}\right)^{2^n}\exp\left(\alpha(\sqrt{2}+1)^2 2^{n+1} y_0^2\right)\right].$$

The last expression is finite provided that the sum of

$$\exp\left(2^n\left(\ln\frac{1-q}{q} + 18\alpha y_0^2\right)\right), \quad n = 1,2,\ldots$$

converges. It converges as long as $\alpha < -\frac{1}{18y_0^2}\ln\frac{1-q}{q} =: \varepsilon$. Since $\frac{1-q}{q} < 1$, ε is strictly positive.

The question arises whether it is it possible to find the exact bound for α, similarly to the finite-dimensional case? The following theorem answers in the affirmative for the supremum-type norms.

Theorem 5.2. *Let the conditions of Theorem 5.1 be fulfilled. Assume in addition that there exists a sequence of linear measurable functionals y_n, $n \geqslant 1$ on $(\mathbb{E},\mathcal{E})$ such that*

$$\mathrm{P}\left(N(X) = \sup_{n\geqslant 1}|(y_n,X)| < \infty\right) = 1.$$

Let us denote $\sigma^2 = \sup_{n\geqslant 1}\mathrm{E}(y_n,X)^2$. Then $\mathrm{E}\exp\alpha N(X)^2 < \infty$ if and only if $\alpha < 1/2\sigma^2$.

We shall prove this theorem in a different but equivalent formulation.

Theorem 5.3. *Let X be a Gaussian zero-mean vector in the space of bounded sequences ℓ^∞ equipped with the Borel σ-algebra $\mathcal{B}_{\ell^\infty}$ and the norm $N(e) = \sup_j|e_j|$. Then $\mathrm{E}\exp\alpha N(X)^2 < \infty$ if and only if $\alpha < 1/2\sigma^2$, where $\sigma^2 = \sup_{n\geqslant 1}\mathrm{E}X_n^2$.*

Jensen's inequality is an important result often used in probability theory.

Lemma 5.4 (Jensen's Inequality). *If a function f is convex, then for any random variable ξ with finite expectation,*

$$\mathrm{E}f(\xi) \geqslant f(\mathrm{E}\xi).$$

Proof. Since f is convex, we have for all x that

$$f(x) \geqslant f(y) + c(x - y),$$

where c only depends on y and, in fact, is equal to $f'(y)$ when the derivative exists. Let us take $x = \xi$, $y = \mathrm{E}\xi$, and apply the expected value operators to both sides. The inequality we obtain is in fact also valid for the case $\mathrm{E}f(\xi) = \infty$.

Proof of Theorem 5.3. For any $h \in (0, \sigma)$ there exists an index $k = k(h)$ such that $\mathrm{E}X^2_{k(h)} \geq (\sigma - h)^2$. Therefore,

$$\mathrm{E}\exp\left(\frac{N(X)^2}{2\sigma^2}\right) \geq \mathrm{E}\exp\left(\frac{X^2_{k(h)}}{2\sigma^2}\right) \geq \left(1 - \frac{(\sigma - h)^2}{\sigma^2}\right)^{-1/2} \to \infty$$

as $h \to 0$.

Now let us prove the inverse implication, that is if $\alpha < 1/2\sigma^2$ then $\mathrm{E}\exp\alpha N(X)^2 < \infty$. To help us here, we orthogonalize the components of the Gaussian vector X. Consider first a finite dimensional vector $\mathbf{X} = (X_1, \ldots, X_n)$. There exists a non-degenerate matrix A that reduces the quadratic form $B(\mathbf{x}) = \mathrm{E}(\mathbf{X}, \mathbf{x})^2$ to the canonical (diagonal) form, $\mathrm{E}(\mathbf{X}, A\mathbf{x})^2 = x_1^2 + \cdots + x_k^2$. Moreover, applying another orthogonalization to $B(\mathbf{x})$, we see that the matrix A can be chosen to be lower-triangular. Therefore the covariance matrix of the vector $A^\top \mathbf{X}$ is diagonal, with the first k diagonal elements being one, and the rest being zeros. Therefore, the first k components $(\lambda_1, \ldots, \lambda_k)$ of $A^\top \mathbf{X}$ are independent standard Gaussian variables, while the other $n - k$ components are zero with probability one. Hence, denoting $\Lambda := (\lambda_1, \ldots, \lambda_k, \lambda_{k+1}, \ldots, \lambda_n)$, where all $\lambda_j, j = 1, \ldots, n$ are independent standard Gaussian variables, we get,

$$\mathbf{X} = (A^\top)^{-1}\Pi_k\Lambda,$$

where Π_k is the projection operator on the first k coordinates, and the matrix $(A^\top)^{-1}\Pi_k$ is lower-triangular. Repeating this orthogonalization successively for $(X_1, \ldots, X_n), n = 2, 3, \ldots$, we obtain that for a Gaussian vector $\mathbf{X}$ in ℓ^∞ there exists a sequence of independent standard Gaussian variables $\lambda_1, \lambda_2, \ldots$ and an infinite lower-triangular matrix B such that

$$X_1 = b_{11}\lambda_1,$$
$$X_2 = b_{12}\lambda_1 + b_{22}\lambda_2,$$
$$\cdots$$
$$X_n = b_{1n}\lambda_1 + b_{2n}\lambda_2 + \cdots + b_{nn}\lambda_n$$
$$\cdots.$$

If we denote the n-th column of B by B_n, then the result above can be rewritten as

$$X = \sum_{n=1}^{\infty} \lambda_n B_n.$$

Now let us consider a sequence of Gaussian random ℓ^∞-valued vectors,

$$X^{-n} = \sum_{k=n}^{\infty} \lambda_k B_k.$$

Notice that by the triangle inequality, with the help of

$$X^{-n} = X - \sum_{k=1}^{n-1} \lambda_k B_k,$$

these vectors indeed lie in ℓ^∞.

Using Theorem 5.1, let us choose the number α for X^{-n+1} and set

$$Y^{-n} = \exp\left(\alpha N(X^{-n})^2\right).$$

Furthermore, let us denote by $\mathcal{A}_{-n}$ the σ-algebra generated by all random variables $\{\lambda_j, j > n\}$. By Jensen's inequality, since the function $\exp(\alpha x^2)$ and the functional $N(e)$ are convex,

$$\begin{aligned}
E\left(Y^{-(n-1)}\middle|\mathcal{A}_{-n}\right) &= E\left(\exp\left(\alpha N\left(\Sigma_{j>n-1}\lambda_j B_j\right)^2\right)\middle|\mathcal{A}_{-n}\right) \\
&\geq \exp\left(\alpha\{E(N(\Sigma_{j>n-1}\lambda_j B_j)|\mathcal{A}_{-n})\}^2\right) \\
&= \exp\left(\alpha\{E(N(\lambda_n B_n + \Sigma_{j>n}\lambda_j B_j)|\mathcal{A}_{-n})\}^2\right) \\
&\geq \exp\left(\alpha\{E(N(\lambda_n B_n)) + N(\Sigma_{j>n}\lambda_j B_j)|\mathcal{A}_{-n})\}^2\right) \\
&= \exp\left(\alpha\left(N\left(\Sigma_{j>n}\lambda_j B_j\right)^2\right)\right) = Y^{-n}.
\end{aligned}$$

The last equality follows from the properties of the conditional expectation. Thus, Y^{-n} is a *submartingale* with the time index running over the negative integers. Notice that, since $EY^{-(n-1)} \geq EY^{-n}$, one can choose α common for all n. Using Kolmogorov-Doob maximal inequality for non-negative submartingales with $Y^0 = \exp(\alpha N(X)^2)$, we have,

$$P(\max_{0\leq j\leq n} Y^{-j} \geq \varepsilon) \leq \varepsilon^{-1} E\exp(\alpha N(X)^2).$$

Since the right-hand side does not depend of n,

$$P(\limsup_{n\to\infty} Y^{-n} \geq \varepsilon) \leq \varepsilon^{-1} E\exp(\alpha N(X)^2).$$

Setting $\varepsilon = \exp(\alpha x^2)$, after some trivial manipulations we obtain that

$$P(\limsup_{n\to\infty} N(X^{-n}) \geq x) \leq e^{-\alpha x^2} E\exp(\alpha N(X)^2).$$

Let us choose x so that the right-hand side is less than one. Since the event $\{\limsup N(X^{-n}) \geq x\}$ belongs to the tail σ-algebra $\bigcap_n \mathcal{A}_{-n}$, its probability can only be equal to 0 or 1. Therefore, with our choice of x, it must equal zero. Thus for any $\varepsilon \in (0, 1/2)$ one can find n such that

$$P(N(\Sigma_{j>n}\lambda_j B_j) \geq x + 1) \leq \varepsilon.$$

Now we repeat the proof of Theorem 5.1 while setting $y_0 = x + 1$ and $q = \varepsilon$. Taking

$$\beta < 18^{-1}(x+1)^{-2}\ln\frac{1-\varepsilon}{\varepsilon},$$

we have,

$$\mathrm{E}\exp\left(\beta N^2\left(\sum_{j>n}\lambda_j B_j\right)\right) < \infty.$$

Let us denote $N_1 = N\left(\sum_{j\leqslant n}\lambda_j B_j\right)$, $N_2 = N\left(\sum_{j>n}\lambda_j B_j\right)$. We have, $N(X) \leqslant N_1 + N_2$. Now let us choose an arbitrary $\alpha < \frac{1}{2\sigma^2}$. Since e^{x^2} is convex,

$$\exp(\alpha N^2(X)) \leqslant \exp(\alpha(N_1 + N_2)^2)$$

$$= \exp\left(\left[(1 - \sqrt{\alpha/\beta})\frac{\sqrt{\alpha}}{1 - \sqrt{\alpha/\beta}}N_1 + \sqrt{\alpha/\beta}\sqrt{\beta}N_2\right]^2\right)$$

$$\leqslant (1 - \sqrt{\alpha/\beta})\exp\left(\frac{\alpha}{(1 - \sqrt{\alpha/\beta})^2}N_1^2\right) + \sqrt{\alpha/\beta}\exp(\beta N_2^2),$$

where ε is chosen sufficiently small in order to make sure that $\beta > \alpha$. The expectation of the second summand is finite. If we now denote $\gamma = \alpha(1 - \sqrt{\alpha/\beta})^{-2}$, it follows by the Cauchy-Bunyakovsky inequality that

$$\mathrm{E}\exp(\gamma N_1^2) \leqslant \mathrm{E}\exp\left(\gamma\Sigma_{j\leqslant n}\lambda_j^2\sup_j\Sigma_k B_{jk}^2\right)$$

$$\leqslant \mathrm{E}\exp\left(\gamma\sigma^2\Sigma_{j\leqslant n}\lambda_j^2\right) = \left(\mathrm{E}\exp\left((\gamma\sigma^2\lambda_1^2)\right)\right)^n$$

$$= (1 - \gamma\sigma^2)^{-n/2}.$$

Here the supremum is taken over all segments of length n from all the rows of matrix B. We note that the sum of squares of all elements in any row is at most σ^2. The right-hand side of the last inequality is finite provided γ is sufficiently small, that is, β is large enough so that $(1 - \sqrt{\alpha/\beta})^2 > 2\alpha\sigma^2$. For that we have to take ε sufficiently small. ∎

Remark 5.5. It is a well-known fact, see e.g. Kolmogorov and Fomin [1999], that the supremum norm is general in the sense that any norm in a normed vector space can be represented as a supremum norm, namely, for any $e \in \mathbb{E}$,

$$\|e\| = \sup_{f\in S_{\mathbb{E}*}}|f(e)|,$$

where $S_{\mathbb{E}*}$ is a unit sphere in the dual space.

6

Hilbert Spaces

Let $X(t,\omega)$, $t \in T$, be a Gaussian function on an arbitrary set T, which, as we know, is the same as a Gaussian random vector in $(\mathbb{R}^T, \mathcal{B}^T)$. Let us introduce a natural metric and a topology in the parameter set T associated with X. Recall that, unless otherwise stated, the expected values of all Gaussian variables are assumed to be zero.

Definition 6.1. *Let $X(t,\omega)$ be a random Gaussian function on a set T. The function*

$$d_X(s,t) = \left(E(X(t) - X(s))^2\right)^{1/2}, \quad (s,t) \in T \times T$$

is called the standard deviation function on T generated by X.

Proposition 6.2. *The function $d_X(s,t)$ is a semimetric on T.*

The proof is trivial, as d is non-negative, satisfies the triangle inequality and vanishes at the points (t,t).

Modifying the space T by considering the points s and t such as $d(s,t) = 0$ to be the same makes d a true metric, and T a metric space.

Now let us consider a different space. Since, by definition, a Gaussian variable has finite variance, it is an element of the Hilbert space $L_2(\Omega, \mathcal{F}, P)$ of all random variables with finite variances. Consider the image $X(T)$ of T in $L_2(\Omega, \mathcal{F}, P)$ under the mapping $X : t \to X(t, \cdot)$. Let us denote by K_X the closure of this image. With a view to work with this vector subspace of $L_2(\Omega, \mathcal{F}, P)$ and to study its properties, we construct the following two representations.

The first one is related to the construction of the closure mentioned above. We denote by $\Phi_0(T)$ the set of all functions on T with finite support, i.e.

$$\varphi \in \Phi_0(T) \Leftrightarrow \operatorname{card} \operatorname{supp} \varphi < \infty.$$

The mapping X induces a mapping $\Phi_0(T)$ to K_X via

$$X(\varphi) = \sum_{t \in \text{supp } \varphi} X(t,\omega)\varphi(t) \in K_X, \quad \varphi \in \Phi_0(T),$$

that is, we take linear combinations of finite numbers of $X(t,\omega)$. The Hilbert structure in $L_2(\Omega,\mathcal{F},\mathrm{P})$ generates a pre-Hilbert structure in $\Phi_0(T)$ via the scalar product

$$\langle \varphi,\psi \rangle = EX(\varphi)X(\psi) = \sum_{s,t \in \text{supp } \varphi} r(s,t)\varphi(s)\psi(t),$$

where $r(s,t)$ is the covariance function of X. For example, for $\varepsilon_t(\cdot)$, the indicator function of a point t, we have that $\varepsilon_t \in \Phi_0(T)$, and $\langle \varepsilon_s,\varepsilon_t \rangle = r(s,t)$, $\|\varepsilon_s - \varepsilon_t\| = d(s,t)$.

Before we describe how we complete the set $\Phi_0(T)$, let us recall what completion really means here. To that end we consider the set $\Phi(T)$ of all Cauchy (fundamental) sequences $f = \{\varphi_n, \varphi_n \in \Phi_0(T), n \geqslant 1\}$. Let $g = \{\psi_n, \psi_n \in \Phi_0(T), n \geqslant 1\}$ be another Cauchy sequence. By the Cauchy-Bunyakovsky inequality, the sequence $\langle \varphi_n,\psi_n \rangle$ is a Cauchy sequence as well, and therefore has a limit. Let us then set

$$\langle f,g \rangle = \lim_{n \to \infty} \langle \varphi_n,\psi_n \rangle.$$

To make sure a pre-Hilbert structure is set up, we assume that the sequences $f,g \in \Phi(T)$, for which $\|f - g\| = 0$ represent the same element in this space. While the space $\Phi(T)$ equipped with this structure is a pre-Hilbert space, it is not necessary separable as the metric space (T,d) is not necessarily separable. But at least we have constructed an isomorphism between K_X and $\Phi(T)$ which is, in fact, a realization of the closure of $X(T)$ in the Hilbert space $L_2(\Omega,\mathcal{F},\mathrm{P})$. Let us state this important result as a proposition.

Proposition 6.3. *The space K_X is isomorphic to the completion $\Phi(T)$ of the set of Gaussian random variables*

$$X(\varphi) = \sum_{t \in \text{supp } \varphi} X(t,\omega)\varphi(t), \quad \varphi \in \Phi_0(T) . \tag{6.1}$$

The restriction of this isomorphism on the set $\Phi_0(T)$ is given by (6.1). The scalar product in $\Phi(T)$ is given by

$$\langle f,g \rangle = EX(f)X(g), \quad \|f\|^2 = EX(f)^2.$$

Now let us construct another representation of K_X, a representation that is more closely connected to the covariance structure of the Gaussian function $X(t,\omega)$, $t \in T$. We map a function φ from $\Phi_0(T)$ to a real function on T, via

$$r : \Phi_0(T) \to \mathbb{R}^T, \quad r_\varphi(\cdot) = \sum_{t \in \text{supp } \varphi} r(\cdot,t)\varphi(t) \in \mathbb{R}^T.$$

In particular, $r_{\varepsilon_t}(\cdot) = r(\cdot, t)$. Moreover,

$$\langle \varphi, \psi \rangle = \sum_{s \in \text{supp } \psi} r_\varphi(s) \psi(s) = \sum_{s \in \text{supp } \varphi} r_\psi(s) \varphi(s).$$

For example, we have that $r_\varphi(t) - r_\psi(t) = \langle \varphi - \psi, \varepsilon_t \rangle$.

Let us denote by $H_0(X)$ the image of $\Phi_0(T)$ in $\mathbb{R}^T$. The pre-Hilbert structure of $\Phi_0(T)$ induces a pre-Hilbert structure in $H_0(X)$ via

$$\langle r_\varphi, r_\psi \rangle = \sum_{s \in \text{supp } \psi} r_\varphi(s) \psi(s) = \langle \varphi, \psi \rangle.$$

In particular, we note that $||r_\varphi||^2 = \sum r(s,t) \varphi(s) \varphi(t) = ||\varphi||^2$.

Proposition 6.4. *For any* φ, ψ *from* $\Phi_0(T)$,

$$\left| \sum r_\varphi(s) \psi(s) \right| \leq ||r_\varphi|| \cdot ||\psi||; \qquad |r_\varphi(t)| \leq ||r_\varphi|| \sqrt{r(t,t)},$$
$$|r_\varphi(t) - r_\varphi(s)| \leq ||r_\varphi|| d(s,t).$$

Proof. The first inequality is nothing but the Cauchy-Bunyakovsky inequality, and the other two follow from the first one by setting $\psi = \varepsilon_t$ and $\psi = \varepsilon_t - \varepsilon_s$, respectively.

Now we can complete the space $H_0(X)$ with its pre-Hilbert structure and get a space $H(X)$, which is then a representation of $K(X)$. From Proposition 6.4 it follows that $H(X)$ is the space of uniformly d-continuous and d-Lipschitz functions on T with the usual topology of uniform convergence.

Now let us consider a complete orthonormal basis $(f_\alpha, \alpha \in A)$ in $H(X)$. Notice that the set A is not necessarily countable and the decomposition

$$f(s) = \sum_{\alpha \in A} f_\alpha(s) \langle f_\alpha, f \rangle$$

means that the sum contains at most a countable number of non-zero terms. Indeed, otherwise the Parseval's identity:

$$\sum_{\alpha \in A} \langle f_\alpha, f \rangle^2 = ||f||^2$$

would not hold.

Theorem 6.5. *Let* $(f_\alpha, \alpha \in A)$ *be a complete orthonormal basis in the space* $H(X)$, *and* $(\varphi_\alpha, \alpha \in A)$ *be the corresponding family of elements from* $\Phi(T)$. *Then the following statements hold.*

a) The family $(X(\varphi_\alpha), \alpha \in A)$ *is a Gaussian orthonormal family.*

b) *For any $t \in T$, the series*

$$\sum_{\alpha \in A} X(\varphi_\alpha) f_\alpha(t) \tag{6.2}$$

converges almost surely, and its sum $\widehat{X}(t, \omega)$ satisfies

$$P(\widehat{X}(t, \omega) = X(t, \omega)) = 1.$$

c) *Let K be a d-semi-compact subset of T, and the trajectories $X(t, \omega)$ are a.s. d-continuous on K. Then, almost surely, the series (6.2) converges uniformly on K, and its sum satisfies,*

$$P(\widehat{X}(t, \omega) = X(t, \omega) \text{ for all } t \in K) = 1.$$

We remind the reader that a semi-compact set is a set such that its arbitrary countable covering with open sets contains a finite sub-covering. It follows then that any semi-compact set contains a countable dense subset.

Proof. The statement **a)** evidently follows from our construction of the Hilbert structure and the isomorphisms of spaces $\Phi(T)$, $H(X)$ and K_X. Indeed,

$$1 = ||f_\alpha||^2 = EX(\varphi_\alpha)^2, \quad 0 = \langle f_\alpha, f_\beta \rangle = EX(\varphi_\alpha) X(\varphi_\beta).$$

Now let us prove statements **b)** and **c)**. First we prove that for any d-semi-compact $K \subset T$, the set of indices A can be chosen to be countable. Indeed, for any t, the series $\sum_{\alpha \in A} f_\alpha(t)^2$ converges absolutely, since the set $(f_\alpha(t), \alpha \in A)$ is the set of the Fourier coefficients of the decomposition of r_{ε_t} over the system $(f_\alpha, \alpha \in A)$. Hence, there exists a countable $A_t \subset A$ such that $f_\alpha(t) = 0$ as $\alpha \notin A_t$. We re-number all f_α so that $\{f_\alpha, \alpha \in A_t\} = \{f_n, n \geqslant 1\}$. Now, rewriting the series $\sum X(\varphi_n) f_n(t)$ as

$$\sum X(\varphi_n) \langle X_t, X(\varphi_n) \rangle_{L_2(\Omega, \mathcal{F}, P)},$$

and using the isomorphism between K_X and $H(X)$, we ensure that this series converges to $X(t, \omega)$ with probability one, so **b)** follows.

Now, since **b)** holds, in view of the Arzela theorem, it is sufficient to show that the partial sums

$$S_n = \sum_{k=1}^{n} X(\varphi_k) f_k$$

are a.s. uniformly bounded and d-equicontinuous on K. Let us choose a countable d-dense subset K_0 of K. Denote $A_K = \cup_{t \in K_0} A_t$. Then for any $\alpha \notin A_K$ we have $f_\alpha(t) = 0$ on K_0, and hence the same holds for any element of K, as $f_\alpha(t)$ is d-continuous on T, being an element of $H(X)$. Now we can re-order all functions f_α, $\alpha \in A_K$ as $(f_n, n \geqslant 1)$.

Let us introduce a sequence of random variables,

$$V_n(\varepsilon) = \sup_{s,t \in K_0, d(s,t) < \varepsilon} |S_n(t) - S_n(s)|,$$

and denote by $\mathcal{F}_n \subset \mathcal{F}$ the σ-algebra generated by random variables $\{ X(\varphi_k), k \leqslant n\}$. We want to prove that $(D_n, \mathcal{F}_n)$, $n \geqslant 0$, where $D_n = \exp(\alpha V_n^2(\varepsilon)) - 1$, is a submartingale for some $\alpha > 0$. Indeed, for any s and t, $(S_n(t) - S_n(s), \mathcal{F}_n)$, $n \geqslant 0$, is a martingale, that is,

$$E(S_n(t) - S_n(s)| \mathcal{F}_n) = S_{n-1}(t) - S_{n-1}(s).$$

Since the norm $N(\cdot) = \sup |\cdot|$ and the function $\exp(\alpha x^2) - 1$ are convex, by Jensen's inequality,

$$E\left(\exp(\alpha V_n^2(\varepsilon)) - 1 \big| \mathcal{F}_n\right) \geqslant \exp(\alpha V_{n-1}^2(\varepsilon)) - 1.$$

By Theorem 5.1 on exponential integrability, there exists $\alpha > 0$ such that the expectations of both sides are finite. By our assumptions $X(t, \omega)$ is a.s. continuous on K, and since K is semi-compact, X is a.s. bounded on K. Therefore, since

$$P(\widehat{X}(t, \omega) = X(t, \omega) \text{ for any } t \in K_0) = 1$$

and because of Theorem 5.1, there exists $\alpha > 0$ such that

$$E \exp\left(\alpha \sup_{s,t \in K_0, d(s,t) < \varepsilon} |\widehat{X}(t,\omega) - \widehat{X}(s,\omega)|^2\right) < \infty.$$

Moreover, for the same α, all ED_n, $n \geqslant 1$ are finite $(ED_n \geqslant ED_{n-1})$. Thus, for this α, D_n, $n \geqslant 1$ is a submartingale.

Let us now apply the Doob-Kolmogorov maximal inequality. After simple manipulations we obtain that

$$P(\sup_{k \geqslant 1} V_k(\varepsilon) > C)$$
$$\leqslant \frac{1}{\exp(\alpha C^2) - 1} E\left(\exp\left(\alpha \sup_{s,t \in K_0, d(s,t) < \varepsilon} |\widehat{X}(t,\omega) - \widehat{X}(s,\omega)|^2\right) - 1\right).$$

As we already mentioned, the functions $\widehat{X}(t, \omega)$ and $X(t, \omega)$ coincide a.s. on K_0, hence, by continuity of $X(t, \omega)$ and that of all partial sums $S_n(t)$,

$$P(\sup_{k \geqslant 1} \sup_{s,t \in K, d(s,t) < \varepsilon} |S_k(t) - S_k(s)| > C)$$
$$\leqslant \frac{1}{\exp(\alpha C^2) - 1} E\left(\exp\left(\alpha \sup_{s,t \in K, d(s,t) < \varepsilon} |X(t,\omega) - X(s,\omega)|\right) - 1\right).$$

Since $X(t, \omega)$ is a.s. d-continuous on K, for any integer $p > 0$ there exists $\varepsilon_p > 0$ such that

$$\mathrm{E}\left[\exp\left(\alpha\sup_{s,t\in K, d(s,t)<\varepsilon_p}|X(t,\omega)-X(s,\omega)|\right)-1\right]<1/p^4.$$

Taking $C = C_p = 1/p$, we have,

$$\sum_{p\geq 1}\mathrm{P}(\sup_{k\geq 1}\sup_{s,t\in K, d(s,t)<\varepsilon_p}|S_k(t)-S_k(s)|>1/p)<\infty.$$

By the Borel-Cantelli lemma, only a finite number of the events in the sum can happen, so the family $\{S_n(t),\ n\geq 1\}$ is uniformly bounded and d-equicontinuous.

7

Separability, Measurability and Oscillations

7.1 Separability and Measurability

Recall from the previous lectures that a Gaussian random process $X(t, \omega)$ defined on an arbitrary set T endows T with a natural metric d. Strictly speaking, d is a semi-metric, and it becomes a metric when we consider it defined not on T but on the quotient space of T with respect to the equivalence relation induced by the semi-metric. As these are not lectures on topology but on probability theory, we ignore such subtleties and use the term "metric" in a somewhat loose sense, always keeping in mind the idea of using the appropriate quotient space. Hopefully purists will forgive us.

In the general theory of random processes, one of the more important questions is the question of existence of a separable and measurable modification of a random process. Recall that a random process (function) $\widetilde{X}(t, \omega)$, $t \in T$ is called a modification of $X(t, \omega)$, $t \in T$, if for any t,

$$P(X(t, \omega) = \widetilde{X}(t, \omega)) = 1.$$

Let us now explain how the notion of separability and measurability is defined with respect to a Gaussian random function. Denote by $B(t, \varepsilon)$ the d-ball with the radius ε and the center at t: $B(t, \varepsilon) = \{s \in T : d(t, s) \leqslant \varepsilon\}$.

Definition 7.1. *A Gaussian random function $X = (X(t, \omega), t \in T, \omega \in \Omega)$ is called* separable *if there exist a countable subset S of T, called a separant, and an event $N_S \in \mathcal{F}$ of zero probability such that for any $\omega \in \Omega \backslash N_S$, $t \in T$, $\varepsilon > 0$, the closure of the set*

$$\{X(s, \omega), s \in S \cap B(t, \varepsilon)\}$$

in $\mathbb{R}$ contains the point $\{X(t, \omega)\} \in \mathbb{R}$.

It is worth noting that the notion of separability is somewhat different in the general theory of stochastic processes. It requires that the following equality holds,

$$\overline{\{X(s,\omega), s \in S \cap B(t,\varepsilon)\}} = \{X(s,\omega), s \in B(t,\varepsilon)\}.$$

(The bar over A, i.e. $\bar{A}$, denotes the closure of A.) Thankfully, when (T,d) is separable, both definitions are equivalent.

Definition 7.2. *A Gaussian random function $X = (X(t,\omega), t \in T, \omega \in \Omega)$ is called measurable if the mapping $(\omega,t) \to X(t,\omega)$ of $(\Omega \times T, \mathcal{F} \otimes \mathcal{J})$ to $(\mathbb{R}, \mathcal{B}_{\overline{\mathbb{R}}})$ is measurable, where $\mathcal{J}$ is the σ-algebra of the subsets of T generated by the set of all d-balls.*

It turns out that the separability of the Gaussian process X and the separability of the metric space (T,d) are closely related.

Theorem 7.3. *Let $X = (X(t,\omega), t \in T, \omega \in \Omega)$ be a Gaussian random function, and let d be the corresponding semi-metric in T. Then the following statements are equivalent.*

a) (T,d) is separable.
b) The Hilbert space $H(X)$ (see Lecture 6) is separable.
c) There exists a separable Gaussian random function $\widehat{X}$ on T such that
    $d_X(s,t) = d_{\widehat{X}}(s,t)$ *for all $s,t \in T$.*
d) There exists a separable and measurable modification $\widetilde{X}$ of X.

Proof. Assertions **d)$\Rightarrow$c)$\Rightarrow$b)$\Rightarrow$a)** are immediate consequences of the separability definition and the representation of the space K_X constructed in Lecture 6, and readers should try to prove them themselves. To prove that **a)** follows **d)**, let S be a countable dense subset of (T,d) such that the distance between any two points from S is strictly positive. Let us number all the points from S and identify each point $s \in S$ with its number. For any $t \in S$ we denote

$$C(t,2^{-n}) = B(t,2^{-n}) \cap \Big(\bigcup_{s \in S, s < t} B(s,2^{-n})\Big)^C,$$

with the notation $(\cdot)^C$ for the complement. It is easy to see that for any $t \in S$ and n,

$$C(t,2^{-n}) \in \mathcal{J}, \quad \bigcup_{t \in S} C(t,2^{-n}) = T, \quad \text{and} \quad C(s,2^{-n}) \cap C(t,2^{-n}) = \emptyset \ \text{ if } \ s \neq t,$$

that is, there exists a measurable partition of T. Now let us consider Gaussian stepwise functions on T,

$$X_n(t,\omega) = X(s,\omega), \quad \text{for all } t \in C(s,2^{-n}), \ s \in S.$$

These functions are obviously measurable. By Chebyshev's inequality,

$$\sum_{n \geqslant 1} P(|X_n(t) - X(t)| > 1/n) \leqslant \sum_{n \geqslant 1} n^2 E(X_n(t) - X(t))^2 \leqslant \sum_{n \geqslant 1} n^2 2^{-2n} < \infty.$$

This means that for any t, $X_n(t) \to X(t)$ with probability one. Let us now define a measurable Gaussian function $\widetilde{X}(t, \omega) := \limsup_{n \to \infty} X_n(t, \omega)$. It is clear that $\widetilde{X}(t, \omega) = X(t, \omega)$ for any t and almost all ω. Now we prove that $\widetilde{X}$ is separable. The set S is countable and the distances between points of S are all positive and $\widetilde{X}(t, \omega)$ has been defined only once at any point, so $P(\widetilde{X}(t, \omega) = X(t, \omega), \ t \in S) = 1$. Let us denote by $s_n(t)$ the point $s \in S$ such that $t \in C(s, 2^{-n})$. For any $\omega \in \Omega$ we obtain that

$$\widetilde{X}(t, \omega) := \limsup_{n \to \infty} X_n(t, \omega) = \limsup_{n \to \infty} X(s_n(t), \omega) = \limsup_{n \to \infty} \widetilde{X}(s_n(t), \omega)$$

and, moreover, $d(s_n(t), t) < 2^{-n}$. Therefore

$$\widetilde{X}(t, \omega) \in \overline{\{\widetilde{X}(s_n(t), \omega), n \geqslant 1, d(s_n(t), t) < \varepsilon\}} = \overline{\{\widetilde{X}(s, \omega), s \in S \cap B(t, \varepsilon)\}},$$

and the assertion follows.

As already mentioned, our definition of separability differs slightly from the classic one; however, all classic properties of separable processes still hold true. For example, the following holds.

Proposition 7.4. *Any set dense in (T, d) is a separant for any d-separable modification of X.*

7.2 Oscillations

In this section we study the set of discontinuity points of a Gaussian random function. For a general random process this set can be arbitrary. Interestingly, for a Gaussian process it is not true at all.

Definition 7.5. *Let (T, δ) be a metric space, f be an $\overline{\mathbb{R}}$-valued function on T. A function*

$$w_f(t) = \lim_{\varepsilon \to 0} \sup_{s, s' \in B(t, \varepsilon)} |f(s) - f(s')|, \ t \in T,$$

is called the δ-oscillation of f. The function

$$v_f(t, u) = \lim_{\varepsilon \to 0} \sup_{s, s' \in B(t, u), \delta(s, s') < \varepsilon} |f(s) - f(s')|.$$

is called the δ-oscillation of f over the ball $B(t, u)$.

Proposition 7.6. *The following statements hold.*

a) $v_f(t,u)$ *is convex in* f *and non-decreasing in* u.
b) If f *is uniformly continuous on* (T,δ), *then* $v_f(t,u) \equiv 0$, *and for any function* q, $v_{f+q}(t,u) = v_q(t,u)$.
c) If $\delta(t,t') < \eta$, *then* $v_f(t,u) < v_f(t,u+\eta)$ *and, moreover,*

$$\lim_{\varepsilon \downarrow 0} v_f(t,u-\varepsilon) \leqslant \liminf_{t' \to t} v_f(t',u);$$

$$\lim_{\varepsilon \downarrow 0} v_f(t,u+\varepsilon) \geqslant \limsup_{t' \to t} v_f(t',u).$$

d) $w_f(t) = \lim_{u \downarrow 0} v_f(t,u)$.
e) $w_f(t) = \limsup_{s \to t}(f_s - f_t) + \limsup_{s \to t}(f_t - f_s)$.

Proofs of these statements are rather simple and follow from the definition of oscillations and the convexity of the uniform norm.

Theorem 7.7 (Itô-Nisio). *Let* (T,δ) *be a metric space, and* $X(t,\omega)$ *be a Gaussian* δ-*separable random function on* T. *Assume that* $d_X(s,t)$ *is uniformly continuous on* (T,δ). *Then there exists an event* $N \in \mathcal{F}$ *of zero probability and an* $\overline{\mathbb{R}}$-*valued function* $\alpha(t)$ *on* T *such that for any* t *and all* $\omega \in \Omega \backslash N$, $w_X(t,\omega) = \alpha(t)$. *In particular this means that the oscillation of* X *is a.s. deterministic.*

Proof. Let us denote by S a countable dense subset of (T,δ). Since the semi-metric d is continuous, S is dense in (T,d) as well. Hence, by Theorem 7.3, S is a separant of X, and $H(X)$ is separable. Let us denote by $\{f_n, n \geqslant 1\}$ an orthonormal basis in $H(X)$, and set $\{\varphi_n, n \geqslant 1\}$ to be the corresponding set of functions on $\Phi(T)$. Let $\mathcal{B}_n$ be the σ-algebra generated by the set of random variables $\{X(\varphi_k), k \geqslant n\}$ (see Lecture 6), augmented with all the events of zero P-probability (simply to be referred to as the P-augmented σ-algebra). Let N_S be the zero P-probability event from Definition 7.1 that defines separability of X. Then for any $s \in S$ and $\omega \in \Omega \backslash N_S$, by Definition 7.1, δ-oscillation of $X(t,\omega)$ over the ball $B(s,u)$, $u > 0$, is equal to the δ-oscillation of $X(t,\omega)$ over $S \cap B(s,u)$. By Theorem 6.5, **b)**, there exists en event N'_S of zero P-probability such that for any $\omega \in \Omega \backslash N'_S$, the functions $X(t,\omega)$ and $\widehat{X}(t,\omega)$ coincide on S. Hence their δ-oscillations over $S \cap B(s,u)$ coincide as well. Since all the functions f_n are uniformly d-continuous, they are also uniformly δ-continuous. Therefore, by Proposition 7.6, **b)**, δ-oscillation of $X(t,\omega)$ over $S \cap B(s,u)$ coincides with δ-oscillation of $\sum_{n+1}^{\infty} X(\varphi_k)(\omega)f_k(t)$ for any n over the same set. Thus for any $s \in S$, $u > 0$ and $\omega \in \Omega \backslash N_S \cap N'_S$, δ-oscillation of $X(t,\omega)$ on $S \cap B(s,u)$ is measurable with respect to the tail σ-algebra $\cap_n \mathcal{B}_n$. Therefore there exists a deterministic function $\alpha(t,u)$ and an event N of zero probability, such that for any $\omega \in \Omega \backslash N$, $t \in S$, $u > 0$, $v_X(t,u) = \alpha(t,u)$. We can define $\alpha(t,u)$ at any point $t \in T$, and include this point in the set S. Since $\alpha(t,u)$ is equal at any point to the δ-oscillation, by Proposition 7.6 **c)** it does not increase in u for any t, and

$$\limsup_{\varepsilon\downarrow 0} \alpha(t, u - \varepsilon) \leqslant \liminf_{s\to t, s\in S} \alpha(s, u) \leqslant \limsup_{s\to t, s\in S} \alpha(s, u) \leqslant \liminf_{\varepsilon\downarrow 0} \alpha(t, u + \varepsilon).$$

Furthermore, since the functions $\alpha(t, u)$ and $w_X(t, u)$ are monotone in u, they are continuous in u in all but a countable number of points, and because of Proposition 7.6 **c)** and the above triple inequality, for any t, they coincide at any point of continuity (at these points $\liminf_{s\to t} \alpha(s, u) = \limsup_{s\to t} \alpha(s, u)$, the same for $v_X(t, u)$). To conclude the proof, let us denote $\alpha(t) = \lim_{u\downarrow 0} \alpha(t, u)$ (the limit does exist for any t, by monotonicity), and recall that, by Proposition 7.6 **d)**, for any $\omega \in \Omega\backslash N$ and all t,

$$w_X(t) = \lim_{u\downarrow 0} v_X(t, u) = \lim_{u\downarrow 0} \alpha(t, u) = \alpha(t).$$

The result we have proven is quite remarkable: for any Gaussian function the set of its discontinuity points is a.s. deterministic. This followed, essentially, from the fact that a Gaussian function can be decomposed into a series of independent random variables. Now we shall see that the sizes of the discontinuity jumps are, in some sense, also deterministic.

Corollary 7.8. *Under the conditions of Theorem 7.7, for any $t \in T$, there exists an event N_t of zero probability such that for all $\omega \in \Omega\backslash N_t$,*

$$\liminf_{s\to t} X(s, \omega) = X(t, \omega) - \frac{1}{2}\alpha(t), \quad \limsup_{s\to t} X(s, \omega) = X(t, \omega) + \frac{1}{2}\alpha(t).$$

Proof. By the Itô-Nisio theorem, for any t, the random variable $\limsup_{s\to t} X(s, \omega) - X(t, \omega)$ is equal to a constant with probability one, a constant that we denote $\beta(t)$. By the symmetry of a Gaussian distribution, the random variable $\limsup_{s\to t}(-X(s, \omega)) - (-X(t, \omega))$ is equal to the same constant almost surely. By Proposition 7.6 **e)**, $\beta(t) = \frac{1}{2}\alpha(t)$.

Corollary 7.9. *Let the conditions of the Itô-Nisio theorem be fulfilled, and furthermore assume that there exists an open set $G \subset T$, a set S dense in G, and a number $a > 0$ such that for any $t \in S$, $\alpha(t) \geqslant a$. Then for any $t \in G$, $\alpha(t) = \infty$.*

Proof. Without loss of generality we can assume that S is countable and is a separant of $X(t, \omega)$ on G. Let N_S be the corresponding zero-probability event and $\omega \in \Omega\backslash(\cup_{s\in S} N_s \cup N \cup N_S)$, where N and N_s are the zero probability events from Theorem 7.7 and Corollary 7.8, respectively. Suppose that for some $t \in G$, $\alpha(t) < \infty$. By the definitions of separability and oscillation, for any $\varepsilon > 0$ one can find points $s, s' \in S$ such that $d(s, t) < \varepsilon/2$, $d(s', t) < \varepsilon/2$ and $X(s, \omega) - X(s', \omega) > \alpha(t) - a/4$. Now let us use Corollary 7.8 and choose two points u, u' from G so that the inequalities

$$d(u, s) < \varepsilon/2, d(u', s') < \varepsilon/2$$

$$X(u,\omega) - X(s,\omega) > \frac{1}{2}\alpha(s) - a/8, \; X(s',\omega) - X(u',\omega) > \frac{1}{2}\alpha(s') - a/8$$

hold. By the triangle inequality, $u, u' \in B(t, \varepsilon)$. Summing up the inequalities for X, it follows from the assumptions that

$$X(u,\omega) - X(u',\omega) > \alpha(t) + \frac{1}{2}\alpha(s) + \frac{1}{2}\alpha(s') - a/2$$

$$\geq \alpha(t) + \frac{1}{2}a + \frac{1}{2}a - \frac{1}{2}a = \alpha(t) + \frac{1}{2}a$$

that in case $\alpha(t) < \infty$ contradicts the Itô-Nisio theorem since $\varepsilon > 0$ can be arbitrarily small.

Corollary 7.9 implies a remarkable fact that if $\alpha(t)$ does not depend of t, it can only be identically equal to zero or to infinity. It equals zero if and only if X is a.s. continuous. Importantly, this is not the case if the process is not Gaussian. The first result of this type, in the case of $X(t,\omega)$ being a stationary Gaussian process on $\mathbb{R}$ (so that $d(s,t)$ depends only on $|s - t|$) was proved by Yu. Belyaev in 1961.

Theorem 7.10 (Belyaev's Alternative). *Let $X(t,\omega)$ be a Gaussian stationary process on an open set $T \subset \mathbb{R}$. Then it is either almost sure continuous or almost sure unbounded on any open sub-interval of T.*

Problem 7.11. Prove that for a Gaussian random function,

$$\alpha(t) = 2 \lim_{\varepsilon \downarrow 0} \mathrm{E} \sup_{s \in B_\delta(t,\varepsilon)} (X(s) - X(t)).$$

The Entropy Method

8.1 Introduction

Let $X(t,\omega)$, $t \in T$, be a Gaussian d-separable random function on T, $EX(t,\omega) \equiv 0$. We again consider the points s,t with $d_X(s,t) = 0$ to be the same so that $d := d_X$ becomes a metric on T. Let us denote by $N(T,\varepsilon)$ the minimal number of d-balls of radius ε covering T :

$$N(T,\varepsilon) = \inf \left\{ n : \exists t_1, \ldots, t_n, \bigcup_{i=1}^{n} B(t_i,\varepsilon) = T \right\}.$$

The set $\{t_1, \ldots, t_n\}$ is called an *ε-net* for T; any ε-net with the smallest possible n is called the *minimal ε-net*; and the quantity $\log_2 N(T,\varepsilon)$ is called *Kolmogorov's ε-entropy* of T or, sometimes, *the entropy complexity* of T. For a metric space (T,d) the quantity $\ln N(T,\varepsilon)$ is often called the *metric entropy* of T.

In 1967, R. Dudley introduced a new tool, the so-called *entropy method*, for studying regularity properties of Gaussian processes. Dudley's integral,

$$\mathcal{D}(u,T) := \int_0^u \sqrt{\ln N(T,x)}dx$$

plays central role here. In this lecture we consider in detail how Dudley's scheme can be applied to estimating exceedance probabilities of Gaussian functions

$$P(\sup_{t \in T} X(t) > u).$$

To start, however, let us consider two remarkable results that have been obtained by this method. The first one is proven by R. Dudley, Dudley [1967].

Theorem 8.1 (Dudley's Theorem). *Let $X(t,\omega)$ with t in an arbitrary metric space (T,d) be a Gaussian d-separable random function with zero* *mean. If for some $h > 0$, $\mathcal{D}(h,T) < \infty$, then $X(t,\omega)$ is a.s. continuous. Moreover,*

$$\mathrm{E}\sup_{t \in T} X(t) \leqslant 4\sqrt{2}\mathcal{D}(\sigma/2, T),$$

where $\sigma^2 = \sup_{t \in T} \mathrm{E}X^2(t)$.

In 1940s, A.N. Kolmogorov posed a very natural problem. Since distributions of Gaussian processes depend only on their means and covariances, can one not come up with the necessary and sufficient conditions for the trajectories to be a.s. continuous in terms of these quantities alone?

The solution to this problem has a long and interesting history, with Dudley's theorem above being one of its concluding phases. X. Fernique, Fernique [1975], gave the complete solution in the case of Gaussian homogeneous fields, that is, Gaussian functions on Euclidean spaces with the constant mean and the covariance function depending only on the difference of its arguments.

Theorem 8.2 (Fernique's Theorem). *Let $X(t,\omega), t \in T \subset \mathbb{R}^d$, be a Gaussian separable homogeneous field. It is a.s. continuous if and only if for some $h > 0$, $\mathcal{D}(h,T) < \infty$.*

We do not prove these theorems here, but illustrate the idea of their proofs with the related problem of bounding from above the exceedance probabilities for a maximum of a Gaussian function.

8.2 The Entropy Method for Estimation of the Exceedance Probability

Lemma 8.3 (V. Dmitrovsky). *Assume that for some $t_0 \in T$, $\mathrm{E}X(t_0,\omega)^2 > 0$, and for some $\delta' > 0$,*

$$\mathcal{D}(1, B(t_0, \delta')) < \infty.$$

Then there exists δ_0 such that for any $\delta < \delta_0$ and all $u > 0$,

$$\mathrm{P}(\sup_{t \in B(t_0,\delta)} X(t) > \sigma u) \leqslant 8\left(1 + \frac{2}{u^2}\right)\overline{\Phi}(u)$$

$$\times \exp\left(2u^2\delta^2\sigma^{-2} + \frac{4\sqrt{2}u}{\sigma}\mathcal{D}(\delta/2, B(t_0,\delta))\right),$$

with $\sigma^2 = \sup_{t \in B(t_0,\delta)} \mathrm{E}X^2(t)$, $\overline{\Phi} = 1-\Phi$, where $\Phi(x)$ is the standard Gaussian distribution function.

Proof. Consider a Gaussian random function

$$\eta(t,\omega) = X(t,\omega) - \mathrm{E}(X(t,\omega)|X(t_0,\omega)).$$

We have learned from Lecture 1 that the random variable $E(X(t,\omega)|X(t_0,\omega))$ is Gaussian, moreover, it can be chosen in such way that

$$E(X(t)|X(t_0)) = \frac{EX(t)X(t_0)}{EX(t_0)^2}X(t_0). \qquad (8.1)$$

We have that $\eta(t_0) = 0$ a.s., $E(\eta(t)|X(t_0)) \equiv 0$ a.s.,

$$E((\eta(t) - \eta(s))^2|X(t_0)) \leqslant d^2(s,t)$$

(The conditional variance is always lower than the unconditional one!) By our assumptions, (T,d) is a separable metric space, therefore the last inequality implies that the Gaussian process η conditioned on $X(t_0)$ can also be chosen to be separable, and this is the modification we shall consider. The last inequality also implies

$$P(\eta(t) - \eta(s) > u|X(t_0)) \leqslant \overline{\Phi}\left(\frac{u}{d(s,t)}\right)$$

and

$$P(\eta(t) > u|X(t_0)) = P(\eta(t) - \eta(t_0) > u|X(t_0)) \leqslant \overline{\Phi}\left(\frac{u}{d(t_0,t)}\right).$$

Furthermore,

$$\sup_{t\in B(t_0,\delta)} \eta(t) \geqslant \sup_{t\in B(t_0,\delta)} X(t) - \sup_{s\in B(t_0,\delta)} E(X(s)|X(t_0)).$$

By Hölder's inequality and the relation (8.1), we have for $t \in B(t_0,\delta)$,

$$E(X(t)|X(t_0)) \leqslant \frac{\sigma}{a}|X(t_0)|, \quad \text{where} \quad a^2 = EX(t_0)^2.$$

Therefore,

$$P(\sup_{t\in B(t_0,\delta)} X(t) > u + \frac{\sigma}{a}|X(t_0)||X(t_0)) \leqslant P(\sup_{t\in B(t_0,\delta)} \eta(t) > u|X(t_0)). \qquad (8.2)$$

Denote $N_k := N(B(t_0,\delta),\delta2^{-k})$ and consider a collection of minimal $\delta2^{-k}$-nets, S_k, $k = 0,1,2,\ldots$, $S_0 = \{t_0\}$, $S_k = \{t_k^i,\ i = 1,...,N_k\}$, on the ball $B(t_0,\delta)$. We associate any point $t \in S_k$ with the set W_k, defined recursively as follows,

$$W_k(t_k^1) = B(t_k^1,\delta2^{-k}),$$
$$W_k(t_k^i) = B(t_k^i,\delta2^{-k})\backslash\bigcup_{j<i} W_k(t_k^j),\ i = 2,\ldots,N_k,$$
$$k = 0,1,2,\ldots.$$

Now let us define a mapping $\theta_k : B(t_0,\delta) \rightarrow S_k$ by the following rule,

$$\theta_k(s) = t \in S_k \;\Leftrightarrow\; s \in W_k(t).$$

We have that $d(s, \theta_k(s)) \leqslant \delta 2^{-k}$ for all $s \in B(t_0, \delta)$. The set $S := \cup_{k=0}^{\infty} S_k$ is d-dense in $B(t_0, \delta)$. Therefore it is a separant for X, η and the chosen modification of the conditional Gaussian process η given $X(t_0)$ simultaneously. For a point $s \in S_n$ there exists a unique sequence of points $\{s_k, k = 0, 1, \ldots, n\}$ such that $s_k \in S_k$, $s_k = \theta_k(s_{k-1})$, and

$$\max_{s \in S(n)} \eta(s) \leqslant \sum_{k=1}^{n} \max_{s \in S_k} (\eta(s) - \eta(\theta_{k-1}s)) \leqslant \sum_{k=1}^{\infty} \max_{s \in S_k} (\eta(s) - \eta(\theta_{k-1}s)).$$

We shall see later that the series on the right-hand side converges. By separability, as the right-hand side does not depend of n, the left-hand side can be replaced with $\sup_{t \in B(t_0, \delta)} \eta(t)$; the resulting inequality holds almost surely. Let us choose $\alpha > 0$ whose value will be specified later. By (8.2),

$$P\left(\sup_{t \in B(t_0, \delta)} X(t) > 2\delta \sum_{k=1}^{\infty} 2^{-k}(u + \sqrt{2 \ln N_k} + \alpha k \ln 2) + \frac{\sigma}{a}|X(t_0)| \,\middle|\, X(t_0) \right)$$

$$\leqslant P\left(\sup_{t \in B(t_0, \delta)} \eta(t) > 2\delta \sum_{k=1}^{\infty} 2^{-k}(u + \sqrt{2 \ln N_k} + \alpha k \ln 2) \,\middle|\, X(t_0) \right)$$

$$\leqslant P\left(\sum_{k=1}^{\infty} \max_{s \in S_k}(\eta(s) - \eta(\theta_{k-1}s)) > 2\delta \sum_{k=1}^{\infty} 2^{-k}(u + \sqrt{2 \ln N_k} + \alpha k \ln 2) \,\middle|\, X(t_0) \right)$$

The event in the last probability contains the event that at least one term from the sum on the left is strictly larger than the term with the same number from the sum on the right, therefore we can continue the chain as

$$\leqslant \sum_{k=1}^{\infty} P\left(\max_{s \in S_k}(\eta(s) - \eta(\theta_{k-1}s)) > 2^{-k+1}\delta(u + \sqrt{2 \ln N_k} + \alpha k \ln 2) \,\middle|\, X(t_0) \right)$$

$$\leqslant \sum_{k=1}^{\infty} N_k \max_{s \in S_k} P\left(\eta(s) - \eta(\theta_{k-1}s) > 2^{-k+1}\delta(u + \sqrt{2 \ln N_k} + \alpha k \ln 2) \,\middle|\, X(t_0) \right)$$

$$\leqslant \sum_{k=1}^{\infty} N_k \max_{s \in S_k} \overline{\Phi}\left(\frac{2^{-k+1}\delta(u + \sqrt{2 \ln N_k} + \alpha k \ln 2)}{d(s, \theta_{k-1}s)} \right).$$

Now, since $d(s, \theta_{k-1}s) \leqslant \delta 2^{-k+1}$, we continue as

$$\leqslant \sum_{k=1}^{\infty} N_k \max_{s \in S_k} \overline{\Phi}\left(u + \sqrt{2 \ln N_k} + \alpha k \ln 2\right)$$

$$= \frac{1}{\sqrt{2\pi}} \sum_{k=1}^{\infty} N_k \int_{u+\alpha k \ln 2}^{\infty} \exp\left(-\frac{1}{2}(x + \sqrt{2 \ln N_k})^2\right) dx$$

$$= \frac{1}{\sqrt{2\pi}} \sum_{k=1}^{\infty} \int_{u+\alpha k \ln 2}^{\infty} \exp\left(-\frac{1}{2}x^2 - \sqrt{2 \ln N_k}x\right) dx$$

$$\leqslant \sum_{k=1}^{\infty} \overline{\Phi}\left(u + \alpha k \ln 2\right).$$

Remark 8.4. If we have a sequence of random variables η_k and a sequence of positive numbers u_k with $\sum u_k < \infty$ and $\sum P(|\eta_k| > u_k) < \infty$, then the series $\sum \eta_k$ converges a.s.. Moreover, $P(\sum \eta_k > \sum u_k) \leqslant \sum P(\eta_k > u_k)$. We do not use this result here, but it is useful to know nonetheless. Its proof follows easily from the Borel-Cantelli lemma and the obvious inequality

$$P(\eta_1 + \eta_2 > u_1 + u_2) \leqslant P(\eta_1 > u_1) + P(\eta_2 > u_2).$$

Let us continue the proof. Denoting

$$u' = 2\delta \sum_{k=1}^{\infty} 2^{-k}(u + \sqrt{2 \ln N_k} + \alpha k \ln 2) + \frac{\sigma}{a}|X(t_0)|$$

$$= 2\delta u + 2\delta \sum_{k=1}^{\infty} 2^{-k}\sqrt{2 \ln N_k} + 4\delta\alpha \ln 2 + \frac{\sigma}{a}|X(t_0)|,$$

we continue with the inequality above (dropping the prime in u'),

$$P\left(\sup_{t \in B(t_0,\delta)} X(t) > u|X(t_0)\right)$$

$$\leqslant \sum_{k=1}^{\infty} \overline{\Phi}\left(\frac{u}{2\delta} - \sum_{i=1}^{\infty} 2^{-i}\sqrt{2 \ln N_i} + \alpha(k - 2)\ln 2 - \frac{\sigma}{2\delta a}|X(t_0)|\right).$$

This is valid for all $\alpha > 0$. Let us denote

$$A = 2\delta \sum_{i=1}^{\infty} 2^{-i}\sqrt{2 \ln N_i},$$

and set

$$\alpha = \frac{2\delta}{2\delta(u - A - \frac{\sigma}{a}|X(t_0)|)}.$$

First suppose that $u - A - \frac{\sigma}{a}|X(t_0)| > 0$. Then

$$P\left(\sup_{t\in B(t_0,\delta)} X(t) > u | X(t_0)\right)$$

$$\leq \sum_{k=1}^{\infty} \overline{\Phi}\left(\frac{1}{2\delta}\left(u - A - \frac{\sigma}{a}|X(t_0)| + \frac{2\delta(k-2)\ln 2}{u - A - \frac{\sigma}{a}|X(t_0)|}\right)\right)$$

$$= \sum_{k=1}^{\infty} \overline{\Phi}\left(\frac{u - A - \frac{\sigma}{a}|X(t_0)|}{2\delta} + \frac{(k-2)\ln 2}{u - A - \frac{\sigma}{a}|X(t_0)|}\right)$$

$$\leq \overline{\Phi}\left(\frac{u - A - \frac{\sigma}{a}|X(t_0)|}{2\delta}\right)\sum_{k=1}^{\infty} \exp(-(k-2)\ln 2)$$

$$= 4\overline{\Phi}\left(\frac{u - A - \frac{\sigma}{a}|X(t_0)|}{2\delta}\right).$$

In the case $2\delta(u - A - \frac{\sigma}{a}|X(t_0)|) \leq 0$, this inequality is evident. By the total probability formula, for sufficiently small δ (as δ_0 can be arbitrarily small),

$$P\left(\sup_{t\in B(t_0,\delta)} X(t) > u\right)$$

$$= \frac{1}{\sqrt{2\pi}a}\int_{-\infty}^{+\infty} e^{-\frac{x^2}{2a^2}} P\left(\sup_{t\in B(t_0,\delta)} X(t) > u | X(t_0) = x\right) dx$$

$$\leq \frac{4}{\sqrt{2\pi}a}\int_{-\infty}^{+\infty} e^{-\frac{x^2}{2a^2}} \overline{\Phi}\left(\frac{u - A - \frac{\sigma}{a}|x|}{2\delta}\right) dx$$

$$= \frac{8}{\sqrt{2\pi}a}\int_{0}^{+\infty} e^{-\frac{x^2}{2a^2}} \overline{\Phi}\left(\frac{u - A - \frac{\sigma}{a}x}{2\delta}\right) dx$$

$$\leq \frac{8}{\sqrt{2\pi}a}\int_{-\infty}^{+\infty} e^{-\frac{x^2}{2a^2}} \overline{\Phi}\left(\frac{u - A - \frac{\sigma}{a}x}{2\delta}\right) dx$$

$$= \frac{8}{\sqrt{2\pi}\sigma}\int_{-\infty}^{+\infty} e^{-\frac{x^2}{2\sigma^2}} \overline{\Phi}\left(\frac{u - A - x}{2\delta}\right) dx = 8\overline{\Phi}\left(\frac{u - A}{\sqrt{\sigma^2 + 4\delta^2}}\right)$$

$$\leq 8\overline{\Phi}\left(\frac{u}{\sigma} - \frac{2u\delta^2}{\sigma^3} - \frac{A}{\sigma}\right) \leq 8\left(1 + \frac{2\sigma^2}{u^2}\right)\overline{\Phi}(u)\exp\left(\frac{2u^2\delta^2}{\sigma^4} + A\frac{u}{\sigma^2}\right).$$

Here we have used the inequality

$$\overline{\Phi}(x - b) \leq \overline{\Phi}(x) e^{bx}\left(1 + \frac{2}{x^2}\right), \quad x, b > 0.$$

which can be easily proved by the reader. As a hint, it follows from the inequalities $\overline{\Phi}(x) \geq \varphi(x)\left(\frac{1}{x} - \frac{1}{x^3}\right)$ and $\overline{\Phi}(x) \leq \frac{1}{x}\varphi(x)$, considering separately the cases $x > b$ and $x \leq b$). It is easy to see also that

$$A \leq 4\sqrt{2}\mathcal{D}(\delta/2, B(t_0, \delta')) < \infty.$$

which concludes the proof.

Theorem 8.5 (V. Dmitrovsky). *Let $X(t,\omega)$ be a Gaussian d-separable process on a set T. Assume that*

$$\mathcal{D}(1,T) < \infty.$$

Then, for any $T_1 \subset T$ and all $u > 0$,

$$P(\sup_{t \in T_1} X(t,\omega) > \sigma u) \leqslant 8\left(1 + \frac{2}{u^2}\right)\overline{\Phi}(u)\exp\left(\chi\left(T_1, \frac{u}{\sigma}\right)\right),$$

where $\sigma^2 = \sup_{t \in T_1} EX^2(t)$,

$$\chi(T_1,u) = \inf_{\varepsilon > 0}\left(2u^2\varepsilon^2 + \ln N(T_1,\varepsilon) + 4\sqrt{2}u \sup_{t \in T_1} \mathcal{D}(\varepsilon/2, T_1 \cap B(t,\varepsilon))\right).$$

Proof. Let S be the minimal ε-net on T_1. Then

$$\sup_{t \in T_1} X(t) = \max_{t \in S} \sup_{s \in T_1 \cap B(t,\varepsilon)} X(s).$$

Hence,

$$P(\sup_{t \in T_1} X(t) > \sigma u) \leqslant N(T_1,\varepsilon)\max_{t \in S} P(\sup_{s \in T_1 \cap B(t,\varepsilon)} X(s) > \sigma u)$$

$$\leqslant 8\left(1 + \frac{2}{u^2}\right)\overline{\Phi}(u)$$

$$\times \exp\left(\frac{2u^2\varepsilon^2}{\sigma^2} + \ln N(T_1,\varepsilon) + \frac{4\sqrt{2}u}{\sigma}\sup_{t \in T_1}\mathcal{D}(\varepsilon/2, T_1 \cap B(t,\varepsilon))\right).$$

To conclude the proof we choose ε to minimize the right-hand side.

By estimating the functional χ in particular cases, one can get fairly exact inequalities for the tails of the distributions of the supremum.

Corollary 8.6. *Under the conditions of Theorem 8.5, for all $u \geqslant u_0$ with*

$$u_0 := \sqrt{2}\max(u : \mathcal{D}(1/\sqrt{u}, T_1)),$$

$$P(\sup_{t \in T_1} X(t,\omega) > \sigma u) \leqslant 8\left(1 + \frac{2}{u^2}\right)\overline{\Phi}(u)\exp\left(6\sqrt{2}(u/\sigma)\mathcal{D}(1/\sqrt{u}, T_1)\right).$$

Proof. Let for $u \geqslant u_0$,

$$\varepsilon(u) = \left(\frac{1}{\sqrt{2u}}\mathcal{D}(1/\sqrt{u}, T_1)\right)^{1/2}.$$

Since $\sqrt{x}$ increases,

$$\chi\,(T_1,u) \leqslant 2u^2\varepsilon^2\,(u) + \ln N(T_1,\varepsilon(u)) + 4\sqrt{2}u\mathcal{D}(\varepsilon(u),T_1)$$

$$\leqslant 2u^2\varepsilon^2\,(u) + \left(\frac{\sqrt{2}}{\varepsilon(u)}\mathcal{D}(\varepsilon(u),T_1)\right)^2 + 4u\sqrt{2}\mathcal{D}(\varepsilon(u),T_1)$$

$$\leqslant 6\sqrt{2}u\mathcal{D}(1/\sqrt{u},T_1).$$

Here we substituted the value of $\varepsilon(u)$ and took into account the fact that under the condition

$$\mathcal{D}(1/\sqrt{u},T_1) \leqslant \sqrt{2} \quad \text{for} \quad u \geqslant u_0,$$

$\varepsilon(u) \leqslant 1/\sqrt{u}$.

From this result and the trivial lower bound $\sup_{t\in T_1} X(t,\omega) \geqslant X(s)$ for any $s \in T_1$, the following asymptotic equality follows,

Corollary 8.7. *Under the conditions of Theorem 8.5,*

$$P(\sup_{t\in T_1} X(t,\omega) > \sigma u) = \overline{\Phi}\,(u)\exp(o(u)), \quad u \to \infty.$$

9

High Excursion Probabilities for Stationary Processes

This lecture starts the part of the course that considers various asymptotic methods for distributions of high maxima of Gaussian processes. Here we are interesting in the behavior of the probability

$$P(\sup_{t \in T} X(t) > u) \tag{9.1}$$

as $u \to \infty$. We shall see that, in a certain sense, these methods are analogous to the Laplace saddle-point approximation method. In the first stage of the approximation we usually look for the most informative domain of integration that is hopefully small enough, although in our case it lives in an infinite-dimensional space. This set gives the main part of the required asymptotic behavior, and we can apply various methods of local analysis within this domain to refine the results.

In 1968 James Pickands III suggested a natural and elegant tool for evaluating the asymptotic behavior of (9.1). Later it was realized that this approach can be applied to a much wider class of Gaussian processes and fields, as well as other Gaussian functions. Before considering various extensions, let us first look at the approach in the original context considered by Pickands.

9.1 Local Lemma

Let $X(t)$, $t \in \mathbb{R}$, be an a.s. continuous Gaussian stationary process, and let us assume that $EX(0) = 0$, $EX^2(0) = 1$. We shall see later on that these restrictions are not important for the essence of the method to be applied. Let us denote by $r(t)$ the covariance function, $r(t - s) = EX(s)X(t)$. Let us assume that for some C and some $\alpha > 0$,

$$r(t) = 1 - C|t|^\alpha + o(|t|^\alpha), \quad t \to 0. \tag{9.2}$$

Clearly any positive C can be changed to $C = 1$ by a time-change in (9.1), so that from now on we consider $C = 1$.

It should be easy for a reader to prove that $\alpha \leqslant 2$, as it follows from the properties of positive-definite functions. From (9.2) it follows that there exists $T > 0$ such that

$$r(t) < 1 \quad \text{for all } t \in (0, T]. \tag{9.3}$$

Sometimes it is useful to make this an explicit assumption, when for example studying the probability 9.1 for large T. This restriction, however, is not essential for the method that we are developing.

Let us set $\chi(t)$ to be equal to $\sqrt{2}B_{\alpha/2}(t) - |t|^\alpha$, where $B_{\alpha/2}(t)$ is the *fractional Brownian motion* with the Hurst exponent equal to $\alpha/2$. That is, let $\chi(t)$ be a Gaussian random process with continuous trajectories, the expected value

$$\mathrm{E}\chi(t) = -|t|^\alpha,$$

and the covariance function

$$r_\chi(s, t) = |s|^\alpha + |t|^\alpha - |t - s|^\alpha.$$

Let us also introduce a function that appears in a number of asymptotic results

$$\Psi(x) = \frac{1}{\sqrt{2\pi x}} e^{-\frac{x^2}{2}}, \quad x > 0. \tag{9.4}$$

Recall from Proposition 2.10 that

$$1 - \Phi(x) \sim \Psi(x)$$

for large x, where Φ is the cumulative distributiobn function of a standard Gaussian random variable.

Lemma 9.1 (Pickands' Lemma). *Under the assumptions listed above, for any $\Lambda > 0$,*

$$\mathrm{P}(\max_{t \in [0, \Lambda u^{-2/\alpha}]} X(t) > u) = H_\alpha(\Lambda)\Psi(u)(1 + o(1)) \quad as \quad u \to \infty,$$

where

$$H_\alpha(\Lambda) = \mathrm{E}\exp\left(\max_{t \in [0, \Lambda]} \chi(t)\right) < \infty.$$

9.1.1 The Beginning of the Proof of Lemma 9.1

Let $u > 0$. By the total probability formula,

$$P(\max_{t\in[0,\Lambda u^{-2/\alpha}]} X(t) > u)$$

$$= \frac{1}{\sqrt{2\pi}} \int_{-\infty}^{\infty} e^{-v^2/2} P\left(\max_{t\in[0,\Lambda u^{-2/\alpha}]} X(t) > u \middle| X(0) = v\right) dv$$

$$= \frac{1}{\sqrt{2\pi}u} e^{-u^2/2} \int_{-\infty}^{\infty} e^{w-w^2/2u^2} P\left(\max_{t\in[0,\Lambda u^{-2/\alpha}]} X(t) > u \middle| X(0) = u - \frac{w}{u}\right) dw,$$

$$(9.5)$$

where we have changed variables $v = u - w/u$. Let us introduce a Gaussian process $\chi_u(t) = u(X(u^{-2/\alpha}t) - u) + w$. The last integral can be rewritten as

$$\int_{-\infty}^{\infty} e^{w-w^2/2u^2} P\left(\max_{t\in[0,\Lambda]} \chi_u(t) > w \middle| X(0) = u - \frac{w}{u}\right) dw. \qquad (9.6)$$

Now let us consider the family of Gaussian distributions that appear in the integral. Let us first look at the corresponding conditional mean and the covariance function. Using formulas from Lecture 1,

$$E\left(\chi_u(t) \middle| X(0) = u - \frac{w}{u}\right) = u\left(E\left(X(u^{-2/\alpha}t) \middle| X(0) = u - \frac{w}{u}\right) - u\right)$$
$$= -u^2(1 - r(u^{-2/\alpha}t)) + w(1 - r(u^{-2/\alpha}t)), \qquad (9.7)$$

$$\text{Var}\left(\chi_u(t) - \chi_u(s) \middle| X(0) = u - \frac{w}{u}\right)$$
$$= u^2\left(\text{Var}[X(u^{-2/\alpha}t) - X(u^{-2/\alpha}s)] - [r(u^{-2/\alpha}t) - r(u^{-2/\alpha}s)]^2\right). \qquad (9.8)$$

Moreover,

$$E\left(\chi_u(0) \middle| X(0) = u - \frac{w}{u}\right) = E\left(\chi_u^2(0) \middle| X(0) = u - \frac{w}{u}\right) = 0.$$

Using (9.2), as $u \to \infty$,

$$E\left(\chi_u(t) \middle| X(0) = u - \frac{w}{u}\right) = -|t|^\alpha + wo(1),$$

and

$$\text{Var}\left(\chi_u(t) - \chi_u(s) \middle| X(0) = u - \frac{w}{u}\right) = -|t - s|^\alpha + o(1).$$

Let us note that the infinitely small variables on the right-hand side do not depend of w. Thus the conditional finite-dimensional distributions of $\chi_u(t)$ converge to the corresponding distributions of $\chi(t)$. Furthermore, the family of the conditional Gaussian distributions is dense in $\mathcal{B}(C[0,\Lambda])$, the σ-algebra in the space of all continuous functions on $[0,\Lambda]$ generated by the cylindric intervals in $C[0,\Lambda]$, $x(t_i) \in (a_i, b_i)$, $i = 1,\ldots,m$, for all m, t_i, a_i, b_i. Indeed, from (9.7), (9.8) it follows that for any positive u, any w from a bounded set W, any $s,t \in [0,\Lambda]$, and any constant C,

$$E\left((\chi_u(t) - \chi_u(s))^2 \,\Big|\, X(0) = u - \frac{w}{u}\right) \leqslant C|t - s|^\alpha. \qquad (9.9)$$

A reader should check this using (9.2). Moreover, for all sufficiently large u and all w,

$$\left|E\left(\chi_u(t) \,\Big|\, X(0) = u - \frac{w}{u}\right)\right| \leqslant \Lambda^\alpha + \frac{1}{2}w. \qquad (9.10)$$

9.1.2 Digression One: On Weak Convergence of Gaussian Measures

I have assumed so far that the reader is familiar with the basic facts on the weak convergence and compactness of families of probability measures, including those generated by random processes as covered in, say, Billingsley [1999]. Fortunately, for the sequences of Gaussian processes, the compactness and convergence conditions can be simplified significantly, as Gaussian distributions are characterized only by their first- and second-order moments.

Definition 9.2. *We say that a sequence of measures μ_n defined on the Borel σ-algebra $\mathcal{B}(S)$ of a metric space (S, ρ) weakly converges to a measure μ if, for any bounded continuous function f on S,*

$$\int f\, d\mu_n \to \int f\, d\mu \text{ as } n \to \infty.$$

A family of measures $\{\mu_a, \alpha \in A\}$ is called sequentially compact *if from any sequence of measures from this family a weakly converging subsequence can be extracted.*

Definition 9.3. *A family of probability measures $\{\mu_a, \alpha \in A\}$ is called tight if for any $\varepsilon > 0$, one can find a sequentially compact $K_\varepsilon \in \mathcal{B}(S)$ such that*

$$\mu_a(K_\varepsilon) > 1 - \varepsilon \quad \text{for all } \alpha \in A.$$

Theorem 9.4 (Yu. V. Prokhorov). *If a family of probability measures $\{\mu_a, \alpha \in A\}$ is tight, it is sequentially compact. If the metric space (S, ρ) is Polish (separable and complete) then, conversely, any sequentially compact set is tight.*

Now we consider Euclidean spaces.

Theorem 9.5 (Arzelá–Ascoli). *For $f \in C([0,1]^d)$, denote $\Delta(f, \delta) = \sup\{|f(\mathbf{t}) - f(\mathbf{s})| : \mathbf{s}, \mathbf{t} \in [0,1]^d, \|\mathbf{t} - \mathbf{s}\| < \delta\}$, $\delta > 0$. The closure of $K \subset C([0,1]^d)$ is a compact if*

$$\sup_{f \in K} |f(\mathbf{0})| < \infty \quad \text{and} \quad \limsup_{\delta \searrow 0} \sup_{f \in K} \Delta(f, \delta) = 0.$$

The following important result follows from the two theorems above, and is also contained in the monograph Billingsley [1999].

Theorem 9.6. *A family of measures $\{\mu_a, a \in A\}$ on $\mathcal{B}(C([0,1]^d))$ is tight (and, hence, sequentially compact) if and only if,*

1. *for any $\varepsilon > 0$ there exists $M > 0$ such that $\mu_a(f : |f(0)| > M)) < \varepsilon$ for all $a \in A$;*
2. *for any $\varepsilon, v > 0$ there exist $\delta > 0$ such that $\mu_a(f : \Delta(f, \delta) > v) < \varepsilon$ for all $a \in A$.*

The following statement is easier to verify than the condition 2. in the case $d = 1$ and countable $A = \{1, 2, \dots\}$.

2'. For any $\varepsilon, v > 0$ there exists a positive integer N and $n_0(\varepsilon, v, N)$ such that for any $n \geqslant n_0(\varepsilon, v, N)$

$$\sum_{i=1}^{N} \mu_n \left(\sup_{(i-1)/N \leqslant s \leqslant i/N} |f(s) - f((i-1)/N)| > v \right) < \varepsilon.$$

An application of the entropy inequality to the exceedance probability for a Gaussian process leads to the following statement, whose proof is suggested as a useful exercise to the interested reader.

Proposition 9.7. *A family $\{X_a(t), t \in [0,1]^d\}$, $a \in A$, of a.s. continuous Gaussian fields is tight in $\mathcal{B}(C([0,1]^d))$ if for some $C, \beta > 0$,*

$$\sup_{a \in A} E((X_a(t) - X_a(s))^2) \leqslant C \|t - s\|^\beta.$$

9.1.3 Digression Two: A Corollary to Dmitrovsky's Inequality

In this section we obtain a corollary to Dmitrovsky's entropy inequality for Gaussian fields with power-like behavior of their covariances at zero. Let $X(t), t \in \mathbb{R}^d$, be a zero-mean Gaussian field with a.s. continuous trajectories. Assume that for some $\alpha, C > 0$

$$d_X(s, t) \leqslant C \|t - s\|^{\alpha/2}. \tag{9.11}$$

It is easy to calculate that for any compact $S \subset \mathbb{R}^d$,

$$N(S, \varepsilon) \sim |S| \varepsilon^{-2d/\alpha}, \ \varepsilon \to 0,$$

where $|S|$ is the volume of S. Let us consider the behavior of Dudley's integral in Theorem 8.5 as $\varepsilon \to 0$. Taking into account that $|B(0, \varepsilon)| \leqslant C\varepsilon^{2d/\alpha}$, considering that $N(B(0, \varepsilon), x) \leqslant C\varepsilon^{2d/\alpha} x^{-2d/\alpha}$ (for perhaps another C),

$$\mathcal{D}(\varepsilon/2, T_1 \cap B(t,\varepsilon)) \leqslant \int_0^{\varepsilon/2} \sqrt{\ln N(B(0,\varepsilon),x)}dx$$

$$\sim \sqrt{\frac{d}{\alpha}} \int_0^{\varepsilon/2} \sqrt{\ln \varepsilon - \ln x}dx = \sqrt{\frac{d}{\alpha}} \int_0^{\varepsilon/2} \sqrt{-\ln(x/\varepsilon)}dx$$

$$= \varepsilon\sqrt{\frac{d}{\alpha}} \int_0^{1/2} \sqrt{-\ln y}dy \sim C\varepsilon,$$

Now let $\varepsilon = 1/u$ in the entropy inequality, so that

$$\exp\left(\chi(S,u)\right) \leqslant CN(S, 1/u) \leqslant C|S|u^{2d/\alpha},$$

where C does not depend of S. Thus we obtain

Proposition 9.8. *Let the condition (9.11) be fulfilled for an a.s. continuous, zero-mean, Gaussian field $X(t)$, $t \in S \subset \mathbb{R}^d$. Then for all positive u and some C,*

$$P\left(\sup_{t \in S} X(t) > u\right) \leqslant C|S|(u/\sigma)^{2d/\alpha}\Psi(u/\sigma),$$

where $\sigma^2 = \sup_{t \in S} EX^2(t)$.

Here we changed $\overline{\Phi}$ to Ψ using asymptotic behavior of the Gaussian tail distribution. We shall see that this upper bound is asymptotically exact later on.

9.1.4 Conclusion of the Proof of Lemma 9.1

Now, from (9.9) it follows that for any bounded W, the collection of distributions

$$P\left(\chi_u \in (\cdot)| X(0) = u - w/u\right), \ u > 0, \ w \in W \tag{9.12}$$

is sequentially compact in $\mathcal{B}(C([0,\Lambda]))$. Moreover, we have already proven the convergence of the mean and the covariance, as $u \to \infty$ and w takes values from a bounded set. Since the function $f \to \max_{t \in [0,\Lambda]} f(t)$ on $C([0,\Lambda])$ is continuous in the uniform metric, for any w,

$$\lim_{u \to \infty} P\left(\max_{t \in [0,\Lambda]} \chi_u(t) > w \middle| X(0) = u - \frac{w}{u}\right) = P\left(\max_{t \in [0,\Lambda]} \chi(t) > w\right).$$

Now let us turn to the convergence of the family of integrals (9.6). As (9.10) holds, we have for sufficiently large u,

$$P\left(\max_{t \in [0,\Lambda]} \chi_u(t) > w \middle| X(0) = u - \frac{w}{u}\right)$$

$$\leqslant P\left(\max_{t \in [0,\Lambda]}(\chi_u(t) - E\chi_u(t)) > w - \max_{t \in [0,\Lambda]} E\chi_u(t) \middle| X(0) = u - \frac{w}{u}\right)$$

$$\leqslant P\left(\max_{t \in [0,\Lambda]}(\chi_u(t) - E\chi_u(t)) > w - \frac{1}{2}|w| \middle| X(0) = u - \frac{w}{u}\right).$$

For the Gaussian process $\chi_u(t) - E\chi_u(t)$, given $X(0) = u - w/u$, the assumptions of Proposition 9.8 are fulfilled. This follows from (9.8) and the condition (9.2). Thus dominating convergence under the integral follows in (9.5), and so the assertion of Lemma 9.1 is proved since for a non-negative random variable with the property $e^x P(\xi > x) \to 0$ as $x \to \infty$, we have $Ee^\xi = \int e^x P(\xi > x) dx$. ∎

Notice that instead of the interval $[0, \Lambda]$ we could have taken any closed set $L \subset [0, \Lambda]$. So, without changing the proof, we get the following generalization of Lemma 9.1.

Lemma 9.9. *Under the assumptions of Lemma 9.1, for any closed $L \subset [0, \Lambda]$,*

$$P\left(\max_{t \in u^{-2/\alpha} L} X(t) > u \right) = H_\alpha(L)\Psi(u)(1 + o(1)) \quad as \quad u \to \infty,$$

where

$$H_\alpha(L) = E \exp \left(\max_{t \in L} \chi(t) \right) < \infty.$$

The probability $P(\max_T X(t) > u)$ is a sub-additive set function on T. Hence, splitting $[0, \Lambda]$ into intervals of length 1 or less and using the fact that $H_\alpha(\Lambda)$ increases in Λ, we get the following simple but important result.

Proposition 9.10. *For any $\alpha \in (0, 2]$, $H_\alpha(\Lambda) \leqslant \Lambda H_\alpha(1)$.*

9.1.5 Estimating the Probability of a Double Event

Let us formulate an important corollary to Lemma 9.9.

Corollary 9.11. *Under the conditions of Lemma 9.9, for any $\Lambda, \Lambda' > 0$ and $\lambda_0 > \Lambda$,*

$$P\left(\max_{t \in [0, \Lambda u^{-2/\alpha}]} X(t) > u, \max_{t \in [\lambda_0 u^{-2/\alpha}, (\lambda_0 + \Lambda')u^{-2/\alpha}]} X(t) > u \right)$$
$$= H_\alpha(\lambda_0, \Lambda, \Lambda')\Psi(u)(1 + o(1)) \quad as \quad u \to \infty,$$

where

$$H_\alpha(\lambda_0, \Lambda, \Lambda') = \int_{-\infty}^{\infty} e^v P\left(\max_{t \in [0, \Lambda]} \chi(t) > v, \max_{t \in [\lambda_0, \lambda_0 + \Lambda']} \chi(t) > v \right) dv < \infty. \tag{9.13}$$

Proof. It follows that for disjoint sets A, B,

$$P(\max_A X(t) > u, \max_B X(t) > u) = P(\max_A X(t) > u) + P(\max_B X(t) > u)$$
$$- P(\max_{A \cup B} X(t) > u).$$

Applying this initially to X, $A = [0, \Lambda u^{-2/\alpha}]$, $B = [\lambda_0 u^{-2/\alpha}, (\lambda_0 + \Lambda')u^{-2/\alpha}]$, and then, after using Lemma 9.9, to the three terms on the right-hand side, and, finally, to χ, $A = [0, \Lambda]$, $B = [\lambda_0, \lambda_0 + \Lambda']$, we obtain the result.

Now let us formulate, although not prove, a two-dimensional analog of Lemma 9.1. We will need it in the proof of the next statement. The proof of this lemma is a word-by-word repetition of the proof of Lemma 9.1, with obvious changes from the one-dimensional case to the two-dimensional one. The reader is encouraged to do the proof in its entirety.

Let $X(s,t)$, $(s,t) \in \mathbb{R}^2$, $EX(s,t) \equiv 0$, $EX^2(t,s) \equiv 1$, be a Gaussian homogeneous (stationary) field, that is, a field with its covariance function equal to $r(s,t) = EX(s_1, t_1)X(s_1 + s, t_1 + t)$ for any s, t, s_1, t_1. Assume that for some $\alpha_1, \alpha_2 > 0$,

$$r(s,t) = 1 - |s|^{\alpha_1} - |t|^{\alpha_2} + o(|s|^{\alpha_1} + |t|^{\alpha_2}), \quad s,t \to 0. \tag{9.14}$$

Denote $\chi(s,t) = \sqrt{2}(B_{\alpha_1/2}(s) + B_{\alpha_2/2}(t)) - |s|^{\alpha_1} - |t|^{\alpha_2}$, with $B_{\alpha_1/2}(s)$, $B_{\alpha_2/2}(t)$, being two independent fractional Brownian motions with Hurst exponents $\alpha_1/2$ and $\alpha_2/2$, respectively. Notice that $\alpha_i \leqslant 2$, $i = 1, 2$, always. Let us introduce a map of $\mathbb{R}^2$ via

$$g_{\alpha_1, \alpha_2} : (s,t) \to (u^{-2/\alpha_1}s, u^{-2/\alpha_2}t).$$

Lemma 9.12. *Under the conditions set out above, for any closed Λ,*

$$P(\max_{(s,t)\in g_{\alpha_1,\alpha_2}\Lambda} X(s,t) > u) = H_{\alpha_1,\alpha_2}(\Lambda)\Psi(u)(1 + o(1)) \qquad as \quad u \to \infty,$$

with

$$H_{\alpha_1,\alpha_2}(\Lambda) = E \exp\left(\max_{(s,t)\in\Lambda} \chi(s,t)\right) < \infty.$$

The following result also holds, in close analogy to the one-dimensional case.

Proposition 9.13. $H_{\alpha_1,\alpha_2}(\Lambda) \leqslant \|\Lambda\| H_{\alpha_1,\alpha_2}([0,1]^2)$, *where $\|\Lambda\|$ is the number of disjoint unit squares covering Λ.*

Let us now consider the process $X(t)$.

Lemma 9.14. *Let $\varepsilon \in (0, 1/2)$ be such that*

$$1 - \frac{1}{2}|t|^\alpha \geqslant r(t) \geqslant 1 - 2|t|^\alpha$$

for all $t \in [0, \varepsilon]$. (Such ε always exists.) Then there exists $h > 0$ such that for any $\Lambda > 0$, $\lambda_0 > \Lambda$, $u \geqslant u_0 := (2(\lambda_0 + \Lambda)/\varepsilon)^{\alpha/2}$,

$$P(\lambda_0, \Lambda) := P\left(\max_{t\in[0,u^{-2/\alpha}\Lambda]} X(t) > u, \max_{t\in[u^{-2/\alpha}\lambda_0, u^{-2/\alpha}(\lambda_0+\Lambda)]} X(t) > u\right)$$

$$\leqslant h\Psi(u) \exp\left(-\frac{1}{8}(\lambda_0 - \Lambda)^\alpha\right).$$

Proof. Let us consider the Gaussian field $X(s,t) = X(s) + X(t)$. We have

$$P(\lambda_0, \Lambda) \leqslant P\left(\max_{(s,t)\in[0,u^{-2/\alpha}\Lambda]\times[u^{-2/\alpha}\lambda_0, u^{-2/\alpha}(\lambda_0+\Lambda)]} X(s,t) > 2u\right).$$

From the conditions of the lemma, we have for the variance $\sigma^2(s,t) = 2(1 + r(t-s))$ of $X(s,t)$,

$$2 \leqslant 4 - 4|t-s|^\alpha \leqslant \sigma^2(s,t) \leqslant 4 - |t-s|^\alpha \qquad (9.15)$$

Therefore, on the set we consider,

$$\sigma^2(s,t) \leqslant 4 - u^{-2}(\lambda_0 - \Lambda)^\alpha.$$

Let us now look at the normalized field $X^*(s,t) = X(s,t)/\sigma(s,t)$. We have,

$$P\left(\max_{(s,t)\in[0,u^{-2/\alpha}\Lambda]\times[u^{-2/\alpha}\lambda_0, u^{-2/\alpha}(\lambda_0+\Lambda)]} X(s,t) > 2u\right)$$

$$\leqslant P\left(\max_{(s,t)\in[0,u^{-2/\alpha}\Lambda]\times[u^{-2/\alpha}\lambda_0, u^{-2/\alpha}(\lambda_0+\Lambda)]} X^*(s,t) > \frac{2u}{\sqrt{4 - u^{-2}(\lambda_0 - \Lambda)^\alpha}}\right).$$

Using (9.15) and the inequality $(a+b)^2 \leqslant 2(a^2 + b^2)$, one can easily bound the metric generated by this field as such,

$$E\left(X^*(s,t) - X^*(s_1,t_1)\right)^2 \leqslant 16(|t-s|^\alpha + |t_1 - s_1|^\alpha).$$

Let us consider two independent, identically-distributed stationary Gaussian processes $\eta_1(t)$ and $\eta_2(t)$ with zero means and covariance functions $\exp(-16|t|^\alpha)$. One can easily see that the covariance function of the field $X^*(s,t)$, $(s,t) \in [0,u^{-2/\alpha}\Lambda] \times [u^{-2/\alpha}\lambda_0, u^{-2/\alpha}(\lambda_0 + \Lambda)]$, dominates, for all $u \geqslant u_0$, the covariance function

$$\frac{1}{2}(\exp(-16|s|^\alpha) + \exp(-16|t|^\alpha))$$

of the Gaussian homogeneous field

$$\eta(s,t) := \frac{\eta_1(s) + \eta_2(t)}{\sqrt{2}}, \quad (s,t) \in [0,u^{-2/\alpha}\Lambda] \times [u^{-2/\alpha}\lambda_0, u^{-2/\alpha}(\lambda_0 + \Lambda)].$$

By Slepian's inequality,

$$P\left(\max_{(s,t)\in[0,u^{-2/\alpha}\Lambda]\times[u^{-2/\alpha}\lambda_0, u^{-2/\alpha}(\lambda_0+\Lambda)]} X^*(s,t) > \frac{2u}{\sqrt{4 - u^{-2}(\lambda_0 - \Lambda)^\alpha}}\right)$$

$$\leqslant P\left(\max_{(s,t)\in[0,u^{-2/\alpha}\Lambda]\times[u^{-2/\alpha}\lambda_0, u^{-2/\alpha}(\lambda_0+\Lambda)]} \eta(s,t) > \frac{2u}{\sqrt{4 - u^{-2}(\lambda_0 - \Lambda)^\alpha}}\right).$$

Notice that we used Slepian's inequality for finite dimensional vectors. But since the fields under consideration are a.s. continuous, the probabilities above can be approximated by the probabilities of maxima over grids with finite number of points. Applying Slepian's inequality from Lecture 2 and increasing the number of points in the grids unboundedly, we obtain Slepian's inequality in continuous time. Now, using the two-dimensional analog of Lemma 9.1, that is, Lemma 9.12, and the Proposition 9.13, to get that the right-hand side in the inequality above is for all $u \geqslant u_0$ at most

$$C\Lambda^2 \Psi\left(\frac{2u}{\sqrt{4 - u^{-2}(\lambda_0 - \Lambda)^\alpha}}\right) \leqslant C_1 \Lambda^2 \Psi(u) \exp\left(-\frac{1}{8}(\lambda_0 - \Lambda)^\alpha\right).$$

9.2 Pickands' Theorem

Theorem 9.15 (J. Pickands III). *Let (9.2) be fulfilled. Then for any p such that $r(t) < 1$, $t \in (0, p]$,*

$$P(\max_{t \in [0,p]} X(t) > u) = H_\alpha p u^{2/\alpha} \Psi(u)(1 + o(1))$$

as $u \to \infty$, where

$$H_\alpha = \lim_{\Lambda \to \infty} \frac{H_\alpha(\Lambda)}{\Lambda}, \quad and \quad 0 < H_\alpha < \infty.$$

Moreover, p can go to zero as $u \to \infty$ as slowly as $pu^{2/\alpha} \to \infty$. In addition, p also can go to infinity provided the right-hand side of this asymptotic relation tends to zero.

Proof. Let us denote

$$\Delta_k = [ku^{-2/\alpha}\Lambda, (k+1)u^{-2/\alpha}\Lambda], \quad \Lambda > 0, \quad N_t = \left\lfloor \frac{t}{u^{-2/\alpha}\Lambda}\right\rfloor,$$

where $\lfloor \cdot \rfloor$ denotes the integer part of a number. By stationarity,

$$P(\max_{t \in [0,p]} X(t) > u) \leqslant (N_p + 1)P(\max_{t \in \Delta_0} X(t) > u),$$

and by Lemma 9.1,

$$\limsup_{u \to \infty} \frac{P(\max_{t \in [0,p]} X(t) > u)}{pu^{2/\alpha}\Psi(u)} \leqslant \frac{H_\alpha(\Lambda)}{\Lambda}. \tag{9.16}$$

Furthermore, by Bonferroni's inequality, using stationarity again,

$$P(\max_{t \in [0,p]} X(t) > u) \geqslant N_p P(\max_{t \in \Delta_0} X(t) > u)$$

$$- 2\sum_{k=1}^{N_p}(N_p - k)P(\max_{t \in \Delta_0} X(t) > u, \max_{t \in \Delta_k} X(t) > u). \tag{9.17}$$

We call this sum the *double sum*, and denote it by Σ_2. We have,

$$\Sigma_2 \leqslant N_p P(\max_{t \in \Delta_0} X(t) > u, \max_{t \in \Delta_1} X(t) > u)$$

$$+ N_p \sum_{k=2}^{N_{\varepsilon}/4} P(\max_{t \in \Delta_0} X(t) > u, \max_{t \in \Delta_k} X(t) > u)$$

$$+ N_p \sum_{k=N_{\varepsilon}/4+1}^{p} P(\max_{t \in \Delta_0} X(t) > u, \max_{t \in \Delta_k} X(t) > u) \tag{9.18}$$

$$=: A_1 + A_2 + A_3,$$

where ε is taken from Lemma 9.14.

Now let us estimate each of the three terms on the right-hand side. Let us begin with A_3. Choose u large enough so that $\Lambda u^{-2/\alpha} \leqslant \varepsilon/16$. Then the distance between Δ_0 and Δ_k in the third sum is at most $\varepsilon/8$, therefore for the terms of A_3 we have, using stationarity,

$$P(\max_{t \in \Delta_0} X(t) > u, \max_{t \in \Delta_k} X(t) > u) \leqslant P(\max_{(s,t) \in \Delta_0 \times \Delta_k} X(s) + X(t) > 2u)$$

$$\leqslant P(\max_{[0,1] \times [1+\varepsilon/8,2]} X(s) + X(t) > 2u).$$

Let us apply Proposition 9.8 to the last probability. To this end notice that

$$\text{Var}(X(s) + X(t)) = 2 + 2r(t - s) \leqslant 4 - 2\max_{s \geqslant \varepsilon/8}(1 - r(s)) = 4 - 2\delta, \quad \delta > 0.$$

By Proposition 9.8,

$$P(\max_{[0,1] \times [1+\varepsilon/8,2]} X(s) + X(t) > 2u) \leqslant Cu^{2/\alpha-1} \exp\left(-\frac{4u^2}{8 - 4\delta}\right)$$

$$= O\left(u^{2/\alpha-1} \exp(-u^2(1 + \delta)/2)\right),$$

as $u \to \infty$. Since N_p increases as a power of p,

$$A_3 = O(\exp(-u^2(1 + \delta')/2)), \tag{9.19}$$

for any $\delta' \in (0, \delta/2)$. The summation limits in A_2 have been chosen so that Lemma 9.14 is applicable, so

$$\limsup_{u \to \infty} \frac{A_2}{N_p \Psi(u)} \leqslant \limsup_{u \to \infty} \frac{N_{\varepsilon}/2\, h\Psi(u) \sum_{k=2}^{N_{\varepsilon}/2} \exp\left(-((k-1)\Lambda)^\alpha/8\right)}{N_p \Psi(u)}$$

$$\leqslant \frac{\varepsilon h}{2p} \sum_{k-1}^{\infty} \exp\left(-(k\Lambda)^\alpha/8\right) \leqslant C \exp\left(-\Lambda^\alpha/8\right), \tag{9.20}$$

where the constant C depends only of ε, p, α.

Now let us find a bound for A_1. We have,

$$P(\max_{t\in\Delta_0} X(t) > u, \max_{t\in\Delta_1} X(t) > u)$$

$$\leqslant P(\max_{t\in\Delta_0} X(t) > u, \max_{t\in u^{-2/\alpha}[\Lambda+\sqrt{\Lambda},2\Lambda]} X(t) > u) + P(\max_{t\in u^{-2/\alpha}[0,\sqrt{\Lambda}]} X(t) > u)$$

$$\leqslant P(\max_{t\in\Delta_0} X(t) > u, \max_{t\in u^{-2/\alpha}[\Lambda+\sqrt{\Lambda},2\Lambda+\sqrt{\Lambda}]} X(t) > u) + P(\max_{t\in u^{-2/\alpha}[0,\sqrt{\Lambda}]} X(t) > u).$$

The second summand is bound using Lemma 9.1 and Proposition 9.10. For the first one we use Lemma 9.14. We obtain that

$$\limsup_{u\to\infty} \frac{A_1}{N_p\Psi(u)} \leqslant h\exp\left(-\left(\sqrt{\Lambda}\right)^\alpha /8\right) + H_\alpha(\sqrt{\Lambda})$$

$$\leqslant h\exp\left(-\Lambda^{\alpha/2}/8\right) + H_\alpha(1)\sqrt{\Lambda}. \tag{9.21}$$

Now let us put all the bounds together, in particular the estimate (9.16) from above and the estimate from below that follows from (9.17) – (9.21). We obtain that for any $\Lambda_1,\Lambda_2 > 0$,

$$\frac{H_\alpha(\Lambda_1)}{\Lambda_1} \geqslant \limsup_{u\to\infty} \frac{P(\max_{t\in[0,p]} X(t) > u)}{pu^{2/\alpha}\Psi(u)} \geqslant \liminf_{u\to\infty} \frac{P(\max_{t\in[0,p]} X(t) > u)}{pu^{2/\alpha}\Psi(u)}$$

$$\geqslant \frac{H_\alpha(\Lambda_2)}{\Lambda_2} - \frac{2Ch}{\Lambda_2}\exp\left(-\Lambda_2^\alpha\right) - \frac{2h}{\Lambda_2}\exp\left(-\frac{1}{8}\Lambda_2^{\alpha/2}\right) - \frac{H_\alpha(1)}{\sqrt{\Lambda_2}}. \tag{9.22}$$

Assume temporally that

$$\liminf_{\Lambda\to\infty} \frac{H_\alpha(\Lambda)}{\Lambda} > 0. \tag{9.23}$$

Letting Λ_2, then Λ_1 go to infinity in (9.22) and using Proposition 9.10, we get that

$$\infty > \liminf_{\Lambda_1\to\infty} \frac{H_\alpha(\Lambda_1)}{\Lambda_1} \geqslant \limsup_{\Lambda_2\to\infty} \frac{H_\alpha(\Lambda_2)}{\Lambda_2} > 0.$$

It follows that the limit of the ratio $H_\alpha(\Lambda)/\Lambda$, as $\Lambda \to \infty$ exists and it is positive. Hence, by (9.22) the theorem follows.

It remains to prove (9.23). To this end consider the parameter set

$$D = \bigcup_j \Delta_{2j} \cap [0,p]$$

and repeat the steps that lead to the estimates from above (9.16) and from below (9.17) – (9.21) for the probability

$$P(\max_{t\in D} X(t) > u).$$

We obtain

$$\frac{H_\alpha(\Lambda_1)}{2\Lambda_1} \geqslant \frac{H_\alpha(\Lambda_2)}{2\Lambda_2} - \frac{2Ch}{\Lambda_2}\exp\left(-\frac{1}{8}\Lambda_2^\alpha\right) \tag{9.24}$$

The "2" in the denominator is from now having half as many intervals Δ_j in the set D, in the asymptotic sense of course. In addition, the last two terms on the right of (9.22) is absent because D does not contain the neighboring intervals Δ_j. Now choose Λ_2 large enough so the following inequality holds

$$\frac{H_\alpha(1)}{2\Lambda_2} - \frac{2C}{\Lambda_2}\exp\left(-\frac{1}{8}\Lambda_2^\alpha\right) > 0.$$

Since $H_\alpha(\Lambda)$ increases, the right hand part of (9.24) is also positive. Taking the lower limit on the left-hand side, we obtain (9.23).

10

Excursion Probabilities for Non-Stationary Processes

In this lecture we consider a problem that, in some sense, is opposite to the problem of Lecture 9. Namely, we consider a Gaussian non-stationary process with the variance attaining its global maximum at a single point.

Let $X(t)$, $t \in [0,T]$, be a Gaussian a.s. continuous process with zero mean, variance $\sigma^2(t) = EX^2(t)$, and correlation function $r(s,t) := EX(s)X(t)/\sigma(s)\sigma(t)$. We assume that $\sigma^2(t)$ reaches its global maximum at an inner point $t_0 \in (0,T)$. The case when t_0 is a boundary point such as, say, $t_0 = 0$, can be handled in the same way; we will mention the necessary modifications in the course of the proof.

Let us introduce the following assumptions on $X(t)$.

E1 For some positive a, β,

$$\sigma(t) = 1 - a|t - t_0|^\beta (1 + o(1)) \text{ with } t \to t_0.$$

E2 (Local stationarity). For some $\alpha \in (0,2]$,

$$r(s,t) = 1 - |t - s|^\alpha (1 + o(1)) \text{ with } s \to t_0, \ t \to t_0.$$

E3 (Regularity). For some positive γ, G and all s, t,

$$E(X(t) - X(s))^2 \leqslant G|t - s|^\gamma.$$

Theorem 10.1. *Let all the above assumptions on $X(t), t \in [0,T]$, be fulfilled, and let t_0 be an inner point of $[0,T]$. Then*

(i) if $\beta > \alpha$,

$$P(\max_{t \in [0,T]} X(t) > u) = \frac{2H_\alpha \Gamma(1/\beta)}{a^{1/\beta}\beta} u^{2/\alpha - 2/\beta} \Psi(u)(1 + o(1))$$

as $u \to \infty$;

(ii) if $\beta = \alpha$,

$$P(\max_{t\in[0,T]} X(t) > u) = H_\alpha^a \Psi(u)(1 + o(1))$$

as $u \to \infty$, where

$$0 < H_\alpha^a = \lim_{\Lambda\to\infty} H_\alpha^a(\Lambda) < \infty,$$

$$H_\alpha^a(\Lambda) = \mathrm{E}\exp\left(\max_{t\in[-\Lambda,\Lambda]} \chi(t) - a|t|^\alpha\right);$$

(iii) if $\beta < \alpha$,

$$P(\max_{t\in[0,T]} X(t) > u) = \Psi(u)(1 + o(1))$$

as $u \to \infty$.

In case $t_0 = 0$, **(i)** holds after dividing the right-hand side by 2; **(ii)** holds with

$$H_\alpha^{0,a}(\Lambda) = \mathrm{E}\exp\left(\max_{t\in[0,\Lambda]} \chi(t) - a|t|^\alpha\right)$$

instead of $H_\alpha^a(\Lambda)$; and **(iii)** does not change. The same is valid for $t_0 = T$.

The above system of relations between α and β, i.e. between the local behavior of the correlation function at zero and the variance at its maximum point, should be clear enough from a physical point of view. If the correlation is declining faster at zero than the variance at its maximum point, then the case **(i)** holds and the movements of the trajectories are un-affected by the slow movement of the variance, so that the behavior of the high excursion probability is close to the case of the stationary $X(t)$. In the opposite case **(iii)** the variance does not affect the movements of the trajectories, so that the probability of high excursion is simply equivalent to that at the single point t_0. The intermediate case **(ii)** is the phase transition, where the variance competes with the movements of the trajectories. An important example of **(i)** is the Brownian bridge $B(t) := W(t) - tW(1)$, $t \in [0,1]$, for which $\alpha = 1$, $\beta = 2$, $t_0 = 1/2$. The Brownian motion $W(t)$, $t \in [0,1]$, when $\alpha = \beta = 1$, $t_0 = 1$, is an important example of **(ii)**.

Proof of the theorem. By time-change $t \mapsto t - t_0$, we can assume that $t_0 = 0$, and $t \in T := [-t_0, T - t_0]$.

First step: extracting an informative small interval. Let us denote $\delta = u^{-2/\beta} \ln^{2/\beta} u$, $A = T\setminus[-\delta,\delta]$. We have for all sufficiently large u,

$$P(\max_{t\in[-\delta,\delta]} X(t) > u) \leqslant P(\max_{t\in T} X(t) > u) \leqslant P(\max_{t\in[-\delta,\delta]} X(t) > u)$$
$$+ P(\max_{t\in A} X(t) > u).$$

In the boundary case one should change $[-\delta,\delta]$ to $[0,\delta]$. For sufficiently large u,

$$\sigma^2(A) := \sup_{t\in A} \sigma^2(t) \leqslant 1 - \frac{a}{2}\delta^\beta,$$

hence, by Proposition 9.8,

$$P(\max_{t \in A} X(t) > u) \leqslant CTu^{2/\gamma-1} \exp\left(-\frac{u^2}{2 - au^{-2}\ln^2 u}\right).$$

On the other hand, it is easy to see that the right-hand side tends to zero faster than $P(X(0) > u)$, and since $P(\max_{t \in [-\delta,\delta]} X(t) > u) \geqslant P(X(0) > u)$,

$$P(\max_{t \in T} X(t) > u) = P(\max_{t \in [-\delta,\delta]} X(t) > u)(1 + o(1)) \qquad (10.1)$$

as $u \to \infty$. In the case $t_0 = 0$ the same holds for $[0, \delta]$.

Second step: the standard process. Consider a Gaussian process

$$Y(t) = \frac{\xi(t)}{1 + b|t|^\beta}, \quad t \in [-\varepsilon, \varepsilon], \ \varepsilon > 0, \ b > 0,$$

where $\xi(t)$ is a Gaussian stationary process with zero mean and the covariance function $r(t) = \exp(-d|t|^\alpha)$, $d > 0$. Let us evaluate the asymptotic behavior of the probability

$$P(\max_{t \in [-\delta,\delta]} Y(t) > u)$$

as $u \to \infty$. We would like to find lower and upper bounds for the probability (10.1) that are close to each other, based on the asymptotic behavior we have established so far and Slepian's comparison inequality with the appropriate selection of b and d. Recall that in case of the boundary point t_0 we would consider the interval $[0, \delta]$ instead.

1. The case $\alpha < \beta$. Choose some $\kappa \in (\alpha, \beta)$ and denote $\Delta = u^{-2/\kappa}$, $\Delta_k = [k\Delta, (k+1)\Delta]$. Let us introduce the following events,

$$A_k = \{\max_{t \in \Delta_k} \xi(t) > u_k\}, \quad A'_k = \{\max_{t \in \Delta_k} \xi(t) > u'_k\}, \ k \in \mathbb{Z},$$

where

$$u_k = \begin{cases} u(1 + b|(k+1)\Delta|^\beta) & \text{if } k < 0, \\ u(1 + b(k\Delta)^\beta) & \text{if } k \geqslant 0, \end{cases}$$

and

$$u'_k = \begin{cases} u(1 + b((k+1)\Delta)^\beta) & \text{if } k \geqslant 0, \\ u(1 + b|k\Delta|^\beta) & \text{if } k < 0. \end{cases}$$

By Bonferroni's inequality,

$$\sum_{-\delta/\Delta-1 \leqslant k \leqslant \delta/\Delta} P(A_k) \geqslant P(\max_{t \in [-\delta,\delta]} Y(t) > u) \geqslant \sum_{-\delta/\Delta \leqslant k \leqslant \delta/\Delta-1} P(A'_k)$$

$$- \sum_{-\delta/\Delta-1 \leqslant k \leqslant \delta/\Delta} \sum_{-\delta/\Delta \leqslant l \leqslant \delta/\Delta-1, l \neq k} P(A_k A_l). \qquad (10.2)$$

In the boundary case, the single sums on both sides start with zero. By Theorem 9.15,

$$P(A_k) = \Delta H_\alpha u_k^{2/\alpha} \Psi(u_k)(1 + \gamma(u_k)),$$

where $\gamma(x) \to 0$ as $x \to \infty$. Similar relation hold for A_k'. Denoting $\widetilde{\gamma}(u) = \sup_{x \geqslant u} |\gamma(x)|$, we have,

$$\sum_{-\delta/\Delta-1 \leqslant k \leqslant \delta/\Delta} P(A_k) \leqslant (1 + \widetilde{\gamma}(u))\Delta H_\alpha \sum_{-\delta/\Delta-1 \leqslant k \leqslant \delta/\Delta} u_k^{2/\alpha} \Psi(u_k)$$

$$\geqslant (1 - \widetilde{\gamma}(u))\Delta H_\alpha \sum_{-\delta/\Delta-1 \leqslant k \leqslant \delta/\Delta} u_k^{2/\alpha} \Psi(u_k).$$

Furthermore, using the expressions for Ψ, u_k and Δ, and denoting $\Delta_1 = u^{2/\beta-2/\kappa}$, we obtain that

$$\Delta \sum_{0 \leqslant k \leqslant \delta/\Delta} u_k^{2/\alpha} \Psi(u_k) = \frac{1}{\sqrt{2\pi}u} e^{-u^2/2} u^{2/\alpha-2/\beta}$$

$$\times \Delta_1 \sum_{0 \leqslant k \leqslant \delta/\Delta} (1 + bu^{-2}(k\Delta_1)^\beta)^{2/\alpha-1} \exp\left(-b(k\Delta_1)^\beta - \frac{1}{2}b^2 u^{-2}(k\Delta_1)^{2\beta}\right).$$

Notice that $\Delta_1 \to 0$ as $u \to \infty$, and, moreover, $\delta/\Delta_1 = u^{-2/\beta+2/\kappa} \ln^{2/\beta} u \to \infty$ as $u \to \infty$. Therefore, by the dominating convergence theorem, the last sum is equivalent to an integral sum at the points $(x_k = k\Delta_1)$,

$$\sum_{0 \leqslant k \leqslant \delta/\Delta} \exp(-b(x_k)^\beta)\Delta_1 \to \int_0^\infty e^{-bx^\beta} dx = b^{1/\beta} \int_0^\infty e^{-x^\beta} dx$$

$$= b^{1/\beta} \beta^{-1} \Gamma(1/\beta),$$

where the last equality follows by the variable change $x^\beta = y$. Repeating the argument above three more times, we make sure that the last three single sums in (10.2) converge to the same limit. In the boundary case there are two sums, not four.

Now we are in a position to estimate the double sum in (10.2). Let us begin with non-neighboring A_k, A_l. We have,

$$P(A_k A_l) \leqslant P(\max_{s \in \Delta_k} \xi(s) > u, \max_{t \in \Delta_l} \xi(t) > u) \leqslant P(\max_{(s,t) \in \Delta_k \times \Delta_l} \xi(s) + \xi(t) > 2u).$$

The variance of the Gaussian field $\xi(s) + \xi(t)$ is equal to $2 + 2r(t - s)$,

$$\max_{(s,t) \in \Delta_k \times \Delta_l} (2 + 2r(t - s)) \leqslant 4 - 2 \max_{(s,t) \in \Delta_k \times \Delta_l} (1 - r(t - s)) \leqslant 4 - 4du^{-2\alpha/\kappa},$$

where the last inequality holds for all sufficiently large u. The variances of increments of the field are at most $4d(|t - t'|^\alpha + |s - s'|^\alpha)$. Applying Proposition 9.8, we obtain that

$$P\left(\max_{(s,t)\in\Delta_k\times\Delta_l}\xi(s)+\xi(t)>2u\right)$$

$$\leqslant Cu^{-4/\kappa+4/\alpha-1}\exp\left(-\frac{4u^2}{2(4-4du^{-2\alpha/\kappa})}\right)$$

$$\leqslant Cu^{-4/\kappa+4/\alpha-1}\exp\left(-\frac{u^2}{2}\right)\exp\left(-\frac{d}{2}u^{2-2\alpha/\kappa}\right).$$

Since $\kappa>\alpha$, $2-2\alpha/\kappa>0$. Thus the order of each term in the double sum is a power of u, multiplied by $\exp\left(-u^2/2\right)$ times $\exp\left(-u^{2-2\alpha/\kappa}/2\right)$. The order for the number of terms is also a power of u. The order of all the single sums, as has already been shown, is a power of u multiplied by $\exp\left(-u^2/2\right)$. Thus, we can conclude from here that the part of the double sum that is taken over non-neighboring events divided by a single sum exponentially decreases to zero as $u\to\infty$.

Consider now the terms over neighboring events. Take $\kappa_1\in(\alpha,\kappa)$, denote $\Delta'_{k+1}=[(k+1)\Delta+u^{-2/\kappa_1},(k+2)\Delta]$, $\Delta''_{k+1}=[(k+1)\Delta,(k+1)\Delta+u^{-2/\kappa_1}]$, and write,

$$P(A_kA_{k+1})\leqslant P(\max_{s\in\Delta_k}\xi(s)>u_k,\max_{t\in\Delta'_{k+1}}\xi(t)>u_{k+1})+P(\max_{t\in\Delta''_{k+1}}\xi(t)>u_{k+1}).$$

The sum over k probabilities on the right, by the argument above, is also exponentially less than the single sum. The sum of the remaining terms can be estimated as a single sum, but since the lengths of intervals Δ''_k are infinitely smaller than the lengths of Δ_k, the sum itself is infinitely smaller. Thus the statement (i) of Theorem 10.1 is proved for $Y(t)$.

Third step: passage to the process X. For an arbitrary $\varepsilon>0$ and all sufficiently large u, if $t\in[-\delta,\delta]$, by **E1**,

$$\frac{1}{1+(a+\varepsilon)|t|^\beta}\leqslant\sigma(t)\leqslant\frac{1}{1+(a-\varepsilon)|t|^\beta}.$$

From here,

$$P\left(\max_{t\in[-\delta,\delta]}\frac{X(t)}{\sigma(t)(1+(a+\varepsilon)|t|^\beta)}>u\right)\leqslant P(\max_{t\in[-\delta,\delta]}X(t)>u)$$

$$\leqslant P\left(\max_{t\in[-\delta,\delta]}\frac{X(t)}{\sigma(t)(1+(a-\varepsilon)|t|^\beta)}>u\right). \quad (10.3)$$

By **E2**, the covariance function of $X(t)/\sigma(t)$, that is, the correlation function $r(t)$ of $X(t)$, satisfies the inequalities

$$\exp(-(1+\varepsilon)|t|^\alpha)\leqslant r(t)\leqslant\exp(-(1-\varepsilon)|t|^\alpha) \quad (10.4)$$

for all sufficiently large u and the same t. From here, by Slepian's inequality, for any b (we are interesting in $b=a\pm\varepsilon$),

$$P\left(\max_{t\in[-\delta,\delta]} \frac{\xi^-(t)}{(1+b|t|^\beta)} > u\right) \leqslant P\left(\max_{t\in[-\delta,\delta]} \frac{X(t)}{\sigma(t)(1+b|t|^\beta)} > u\right)$$

$$\leqslant P\left(\max_{t\in[-\delta,\delta]} \frac{\xi^+(t)}{(1+b|t|^\beta)} > u\right), \qquad (10.5)$$

where the covariance functions of the zero-mean Gaussian stationary processes $\xi^-(t)$ and $\xi^+(t)$ are equal to the right- and left-hand parts of (10.4) respectively. The asymptotic behavior of the right- and left-hand sides of (10.5) have already been determined. They only differ in the coefficients that can be made arbitrarily close by taking a sufficiently small ε. The statement **(i)** follows.

2. The case $\alpha = \beta$. We need another generalization of Lemma 9.1.

Lemma 10.2. *Let the conditions* **E1** − **E3** *with* $\alpha = \beta$ *hold for a Gaussian process* $X(t)$. *Then for any* $\Lambda_1, \Lambda_2 > 0$,

$$P\left(\max_{t\in[-\Lambda_1 u^{-2.\alpha}, \Lambda_2 u^{-2.\alpha}]} X(t) > u\right) = H_\alpha^a(\Lambda_1, \Lambda_2)\Psi(u)(1 + o(1)), \quad u \to \infty,$$

where

$$H_\alpha^a(\Lambda_1, \Lambda_2) = \mathrm{E}\exp\left(\max_{t\in[-\Lambda_1,\Lambda_2]} \chi(t) - a|t|^\alpha\right)$$

Proof. The proof differs from the proof of Lemma 9.1 only in the calculations of the limits of the conditional mean and variance of the process $\chi_u(t)$. Let us recall the beginning of that proof. Set $u > 0$. By the total probability formula,

$$P(\max_{t\in[0,\Lambda u^{-2/\alpha/}]} X(t) > u) \qquad (10.6)$$

$$= \frac{1}{\sqrt{2\pi}} \int_{-\infty}^{\infty} e^{-v^2/2} P\left(\max_{t\in[-\Lambda_1 u^{-2.\alpha}, \Lambda_2 u^{-2.\alpha}]} X(t) > u \,\Big|\, X(0) = v\right) dv$$

$$= \frac{1}{\sqrt{2\pi u}} e^{-u^2/2} \int_{-\infty}^{\infty} e^{w-w^2/2} P\left(\max_{t\in[-\Lambda_1 u^{-2.\alpha}, \Lambda_2 u^{-2.\alpha}]} > u \,\Big|\, X(0) = u - \frac{w}{u}\right) dw,$$

where we have changed $v = u - w/u$. Let us introduce a Gaussian process $\chi_u(t) = u(X(u^{-2/\alpha}t) - u) + w$, so that the last integral can be rewritten as

$$\int_{-\infty}^{\infty} e^{w-w^2/2} P\left(\max_{t\in[-\Lambda_1,\Lambda_2]} \chi_u(t) > w \,\Big|\, X(0) = u - \frac{w}{u}\right) dw.$$

Now let us consider the family of Gaussian distributions inside the integral. Let us find the limits on the conditional expected value and covariance for this family, and study its tightness. We have, using the formulas that should be well-known by now,

$$E\left(\chi_u(t)|X(0) = u - \frac{w}{u}\right) = u\left(E\left(X(u^{-2/\alpha}t)\middle|X(0) = u - \frac{w}{u}\right) - u\right)$$

$$= u\left(\frac{EX(u^{-2/\alpha}t)X(0)}{EX(0)^2}\left(u - \frac{w}{u}\right)\right)$$

$$= -u^2(1 - \sigma^2(u^{-2/\alpha}t)r(u^{-2/\alpha}t)) + w(1 - \sigma^2(u^{-2/\alpha}t)r(u^{-2/\alpha}t)) \qquad (10.7)$$

and

$$\mathrm{Var}\left(\chi_u(t) - \chi_u(s)|X(0) = u - \frac{w}{u}\right)$$

$$= E\left(\chi_u(t) - \chi_u(s) - E\left(\chi_u(t) - \chi_u(s)|X(0) = u - \frac{w}{u}\right)\right)^2$$

$$= u^2\left(E[X(u^{-2/\alpha}t) - X(u^{-2/\alpha}s)]^2\right.$$

$$\left. - [\sigma^2(u^{-2/\alpha}t)r(u^{-2/\alpha}t) - \sigma^2(u^{-2/\alpha}s)r(u^{-2/\alpha}s)]^2\right). \qquad (10.8)$$

Moreover,

$$E\left(\chi_u(0)|X(0) = u - \frac{w}{u}\right) = E\left(\chi_u^2(0)|X(0) = u + \frac{w}{u}\right) = 0.$$

Now apply **E2**, **E3**. We get that, as $u \to \infty$,

$$E\left(\chi_u(t)|X(0) = u - \frac{w}{u}\right) = -|t|^\alpha - a|t|^\alpha + wo(1),$$

and

$$\mathrm{Var}\left(\chi_u(t) - \chi_u(s)|X(0) = u - \frac{w}{u}\right) = -|t - s|^\alpha + o(1).$$

The rest of the proof repeats the proof of Lemma 9.1 essentially word for word.

Let us continue the proof of the theorem for the case under consideration. Again, we partition $[-\delta, \delta]$ into small intervals Δ_k. Here we come back to the intervals used in the proof of Theorem 9.15, that is, we set $\Delta = \Lambda u^{-2/\alpha}$, $\Delta_k = [k\Delta, (k+1)\Delta]$. For all sufficiently large u,

$$P\left(\max_{t \in [-\delta, \delta]} X(t) > u\right) \geqslant P\left(\max_{t \in [-\Lambda_1 u^{-2/\alpha}, \Lambda_2 u^{-2/\alpha}]} X(t) > u\right).$$

Furthermore,

$$P\left(\max_{t \in [-\delta, \delta]} X(t) > u\right) \leqslant P\left(\max_{t \in [-\Lambda_1 u^{-2/\alpha}, \Lambda_2 u^{-2/\alpha}]} X(t) > u\right)$$

$$+ \sum_{k=-[\Delta/\delta]-1, k \neq 0, -1}^{[\Delta/\delta]+1} P(A_k),$$

where $A_k = \{\max_{\Delta_k} X(t) > u\}$. In the case of the inner point t_0, we take $\Lambda_1 = \Lambda_2 = \Lambda$, and in case of the boundary point, we use $\Lambda_1 = 0$, $\Lambda_2 = \Lambda$. The limits of summation in the case of a boundary t_0 should be changed in an obvious way. Applying (10.3) – (10.5), we have

$$P(A_k) \leqslant P\left(\max_{t \in \Delta_k} \xi(t) > u(1 + |k\Delta|^\alpha)\right) = P\left(\max_{t \in \Delta_k} \xi(t) > u(1 + |\Lambda k|^\alpha u^{-2})\right)$$

which, given Lemma 9.1, is at most $4\Psi(u)H_\alpha(\Lambda)\exp(-|\Lambda k|^\alpha)$ for sufficiently large u and all k. Indeed,

$$P\left(\max_{t \in \Delta_k} \xi(t) > u(1 + |\Lambda k|^\alpha u^{-2}\right)$$

$$\leqslant (1 + \widetilde{\gamma}(u))H_\alpha \Psi\left(u(1 + |\Lambda k|^\alpha u^{-2})\right)$$

$$\leqslant \frac{(1 + \widetilde{\gamma}(u))H_\alpha}{u\sqrt{2\pi}(1 + |\Lambda k|^\alpha u^{-2})} e^{-\frac{1}{2}u^2 - |\Lambda k|^\alpha}$$

$$\leqslant 2H_\alpha(1 + u^{-2})\Psi(u)e^{-|\Lambda k|^\alpha} \leqslant 4H_\alpha\Psi(u)e^{-|\Lambda k|^\alpha},$$

for all $u \geqslant u_0$, where u_0 is the solution to the equation $\max(\widetilde{\gamma}(u), u^{-2}) = 1$ with

$$\widetilde{\gamma}(u) = \sup_{v \geqslant u} \left|\frac{P\left(\max_{t \in \Delta} \xi(t) > v\right)}{H_\alpha \Psi(v)} - 1\right| \downarrow 0 \quad \text{as} \quad u \uparrow \infty.$$

Therefore,

$$\sum_{k=-[\Delta/\delta]-1, k \neq 0, -1}^{[\Delta/\delta]+1} P(A_k) \leqslant 8H_\alpha\Psi(u)\sum_{k=1}^{\infty} e^{-|\Lambda k|^\alpha},$$

that is,

$$\limsup_{u \to \infty} \frac{\sum_{k=-[\Delta/\delta]-1, k \neq 0, -1}^{[\Delta/\delta]+1} P(A_k)}{P\left(\max_{t \in [-\Lambda u^{-2/\alpha}, \Lambda u^{-2/\alpha}]} X(t) > u\right)}$$

$$\leqslant \limsup_{u \to \infty} \frac{8H_\alpha\Psi(u)\sum_{k=1}^{\infty} e^{-|\Lambda k|^\alpha}}{P(X(0) > u)} \to 0$$

as $\Lambda \to \infty$. Now let us take an arbitrarily small $\varepsilon > 0$, and let Λ be as large as the last lim sup above does not exceed ε. Then

$$1 = \lim_{u \to \infty} \frac{P\left(\max_{t \in [-\Lambda_1 u^{-2/\alpha}, \Lambda_2 u^{-2/\alpha}]} X(t) > u\right)}{H_\alpha^a(\Lambda_1, \Lambda_2)\Psi(u)}$$

$$\leqslant \liminf_{u \to \infty} \frac{P\left(\max_{t \in [-\delta, \delta]} X(t) > u\right)}{H_\alpha^a(\Lambda_1, \Lambda_2)\Psi(u)}$$

$$\leqslant \limsup_{u \to \infty} \frac{P\left(\max_{t \in [-\delta, \delta]} X(t) > u\right)}{H_\alpha^a(\Lambda_1, \Lambda_2)\Psi(u)} \leqslant 1 + \varepsilon,$$

where $\Lambda_1 = \Lambda_2 = \Lambda$. For the boundary point t_0, we would have $\Lambda_1 = 0$, $\Lambda_2 = \Lambda$. From here the statement **(ii)** follows.

3. Case $\alpha > \beta$. Notice that in this case $\delta = o(u^{-2/\alpha})$ as $u \to \infty$. Hence, by the arguments of the proof of **(i)**, we obtain that for an arbitrary small $\varepsilon > 0$ and all sufficiently large u,

$$1 = \lim_{u \to \infty} \frac{P(X(0) > u)}{\Psi(u)} \leqslant \liminf_{u \to \infty} \frac{P\left(\max_{t \in [-\delta, \delta]} X(t) > u\right)}{\Psi(u)}$$

$$\leqslant \limsup_{u \to \infty} \frac{P\left(\max_{t \in [-\delta, \delta]} X(t) > u\right)}{\Psi(u)}$$

$$\leqslant \lim_{u \to \infty} \frac{P\left(\max_{t \in [-\varepsilon u^{-2/\alpha}, \varepsilon u^{-2/\alpha}]} \xi(t) > u\right)}{\Psi(u)} = H_\alpha(\varepsilon).$$

By the monotone convergence theorem, $H_\alpha(\varepsilon) \to 1$ as $\varepsilon \to 0$. Thus the statement **(iii)** follows, and the theorem is proved in its entirety.

11

Excursion Probabilities. Examples

In this lecture we apply the results of Lectures 9 and 10 to several well-known Gaussian processes.

11.1 Brownian Motion

The Brownian motion, or the Wiener process, is a Gaussian process W_t, $t \geqslant 0$, that starts at 0, $W_0 = 0$, has zero mean, independent increments, and the following variance of increments,

$$E(W_t - W_s)^2 = \sigma^2(t - s), \quad t \geqslant s.$$

It follows that there exists a continuous modification that we consider from now on. The covariance function is equal to $R(s,t) = \sigma^2 \min(s,t)$. If $\sigma^2 = 1$, the process is called the standard Brownian motion (or the standard Wiener process). Applying Theorem 10.1, we can find the asymptotic behavior of the probability $P(\max_{t \in [0,p]} W_t > u)$ as $u \to \infty$. It may seem a bit pointless as we know the exact distribution of its maximum from the reflection principle: the probability is exactly equal to $2P(W_p > u)$. Nevertheless let us continue, as it would serve as a good warm-up exercise for future investigations of tight asymptotics for a class of similar processes. Furthermore, we will be able to find the value of the constant H in Theorem 10.1. We consider the standard Brownian motion $\sigma = 1$. Now $\sigma^2(t) = \mathrm{Var} W_t = t$, so t_0 where the maximum of the variance is achieved, is given by $t_0 = p$. Furthermore,

$$\sigma(t) = \sqrt{p} - \frac{1}{2\sqrt{p}}|t - p| + o(|t - p|), \quad t \to p$$

(by the Taylor formula for the square root) and, as $t \to p$, $s \to p$,

$$r(s,t) = \frac{\min(s,t)}{\sqrt{st}} = \frac{\frac{1}{2}\left(|s| + |t| - |t - s|\right)}{\sqrt{st}} = \frac{|s| + |t|}{2\sqrt{st}} - \frac{|t - s|}{2\sqrt{st}}$$

$$= 1 + \left(\frac{|s| + |t|}{2\sqrt{st}} - 1\right) - \frac{1}{2p}|t - s| + |t - s|O(|t - p| + |s - p|)$$

$$= 1 + \frac{(\sqrt{t} - \sqrt{s})^2}{2\sqrt{st}} - \frac{1}{2p}|t - s|(1 + O(|t - p| + |s - p|))$$

$$= 1 + O((t - s)^2) - \frac{1}{2p}|t - s|(1 + O(|t - p| + |s - p|))$$

$$= 1 - \frac{1}{2p}|t - s|(1 + o(1)).$$

Now, in order to obtain the constants in the form defined by the conditions of Theorem 10.1, we need to scale the time, $t' = t/2p$, so that

$$P(\max_{t\in[0,p]} W_t > u) = P\left(\frac{\max_{t\in[0,p]} W_t}{\sigma\sqrt{p}} > \frac{u}{\sigma\sqrt{p}}\right)$$

$$= P\left(\max_{t'\in[0,1/2]} \frac{W_{2pt'}}{\sigma\sqrt{p}} > \frac{u}{\sigma\sqrt{p}}\right).$$

Now the process on the right-hand side satisfies the conditions **E1 − E3** and in addition we note that $\alpha = \beta = 1$, $a = 1$. Thus

$$P(\max_{t\in[0,p]} W_t > u) = H_1^{0,1}\Psi(u/\sigma\sqrt{p})(1 + o(1)),$$

where

$$H_1^{0,1} = \lim_{\Lambda\to\infty} E\exp\left(\max_{t\in[0,\Lambda]} \chi(t) - |t|\right).$$

It is easy to see that if $\alpha = 1$, then $\chi(t) = \sqrt{2}W_t - |t|$, that is,

$$H_1^{0,1} = \lim_{\Lambda\to\infty} E\exp\left(\sqrt{2}\max_{t\in[0,\Lambda]} W_t - 2|t|\right).$$

Since the distribution of the maximum of the Brownian motion is known explicitly, we obtain $H_1^{0,1} = 2$. Therefore, for any Gaussian process with the maximum of its variance attained at the boundary point, with the linear behavior of the variance near this point, and the behavior of the correlation function like $1 - |t - s|$, we obtain the asymptotic formula for the excursion probability *with explicit constants*.

11.2 Brownian Bridge

The Brownian bridge $B(t)$, $t \in [0,1]$, is a Gaussian process with zero mean and the covariance function

$$r_B(s,t) = \min(s,t) - st.$$

It is easy to establish the following representation for the Brownian bridge:

$$B(t) = W_t - tW_1,$$

where W is the standard Brownian motion. Conversely, for a standard Gaussian random variable ξ independent of $B(t)$, the Gaussian process $B(t) + \xi t$ is a standard Brownian motion on $[0,1]$. The maximum of its variance $\sigma^2(t) = t - t^2$ is reached at $t_0 = 1/2$. Furthermore,

$$\sigma(t) = \frac{1}{2} - \left(t - \frac{1}{2}\right)^2 (1 + o(1)), \quad t \to \frac{1}{2}.$$

Having in mind that $E(B(t) - B(s))^2 = |t - s| - (t - s)^2$, and $2EB(s)B(t) = EB(s)^2 + EB(t)^2 - E(B(t) - B(s))^2$, we find that, as $s,t \to 1/2$,

$$r(s,t) = 1 - \frac{1}{2}|t - s|(1 + o(1)), \quad t - s \to 0.$$

By Theorem 10.1,

$$P(\max_{t \in [0,1]} B(t) > u) \sim H_1 e^{-2u^2}$$

as $u \to \infty$. Since the distribution of the maximum is known,

$$P(\max_{t \in [0,1]} B(t) > u) = e^{-2u^2},$$

we obtain the value of Pickands' constant for $\alpha = 1$: $H_1 = 1$.

It follows that we can calculate the exact asymptotic behavior of excursion probabilities for Gaussian stationary processes with "Brownian" behavior of its covariance at zero, that is, with $\alpha = 1$. Two examples of such processes are the Ornstein-Uhlenbeck process, with the covariance function $\exp(-\lambda|t|)$, and the process with the "triangle" covariance function: $\max(1 - |at|,0)$ λ, $a > 0$. In order to prove that this function is indeed a valid covariance function, one should calculate the covariance function of $W(t + 1) - W(t)$, first proving that this process is stationary.

11.3 Fractional Brownian Motion

We already know that the fractional Brownian motion $B_H(t)$, $t \geqslant 0, H \in (0,1]$, is a Gaussian process with stationary increments, zero mean, starting at zero, $B_H(0) = 0$, and with the variance of increments

$$E(B_H(t) - B_H(s))^2 = \sigma^2|t - s|^{2H}.$$

If $\sigma^2 = 1$, the process is called the standard fractional Brownian motion. H is called the Hurst exponent. With $H = 1/2$, the process is the Brownian motion. It is interesting to note that if $H = 1$ the process is degenerate so that $B_1(t) = \sigma \xi t$, with ξ being a standard Gaussian variable[1].

Let us again assume for simplicity that $\sigma^2 = 1$, and let us find the asymptotic behavior of $P(\max_{t \in [0,1]} B_H(t) > u)$. The trivial case $H = 1$ is excluded, also for simplicity. We already have enough experience to calculate that as $s, t \to 1$,

$$\sigma(t) = 1 - 2H|1 - t|(1 + o(1))$$

and

$$r(s,t) = 1 - \frac{1}{2}|t - s|^{2H}(1 + o(1)).$$

This means that $\alpha = 2H$, $\beta = 1$. The problem has already been solved for $H = 1/2$. If $H > 1/2$, the condition (iii) is satisfied, and so we have the following result. Recall the function Ψ given by (9.4).

Proposition 11.1. *For $H > 1/2$,*

$$P(\max_{t \in [0,1]} B_H(t) > u) \sim P(B_H(1) > u) \sim \Psi(u)$$

as $u \to \infty$.

If $H < 1/2$, the condition (i) is satisfied, and after the time change $t' = 2^{-1/(2H)}t$, we obtain that $a = 2H2^{-1/(2H)}$.

Proposition 11.2. *For $H < 1/2$*

$$P(\max_{t \in [0,1]} B_H(t) > u) \sim \frac{H2H}{2H} 2^{\frac{1}{2H}} u^{\frac{1}{H}-2} \Psi(u)(1 + o(1))$$

as $u \to \infty$.

11.4 The Ruin Problem for Fractional Brownian Motion

A common problem in actuarial mathematics is the *ruin problem* for an insurance company. Premiums are paid by customers at a constant rate $c > 0$, and insurance claims arrive according to a random step-wise increasing process $S_1(t)$. For an insurer starting with the initial capital x, his surplus at time t is described by the model

$$K(t) = x - S_1(t) + ct,$$

A particular case of this model is known as the Cramer-Lundberg model. The ruin problem here is to estimate the *ruin probability*

---

[1] The reader should prove this, and prove also that if $H > 1$, the process is deterministic.

$$P\left(\exists t \geqslant 0 : K(t) < 0\right).$$

For large x, the problem is solved by Cramer-Lundnerg's theorem that is based on the random walk theory.

If the insurer has other income, say $S_2(t)$, modeled by another step-wise increasing process, the model becomes more complicated:

$$K(t) = x - S_1(t) + S_2(t) + ct.$$

In large insurance companies the jumps in S_1 and S_2 happen very often, so that the sum $-S_1(t) + S_2(t)$ behaves like a continuous random process, with a popular model being the Brownian motion. Sometimes a fractional Brownian motion is used instead since the jumps, in general, could be dependent. As an example of an application of Theorem 10.1 in actuarial mathematics, we consider the model

$$K(t) = x + B_H(t) + ct, \quad t \geqslant 0,$$

with the ruin probability

$$P\left(\exists t \geqslant 0 : K(t) < 0\right) = P\left(\sup_{t \geqslant 0}(-B_H(t) - ct) > x\right)$$

$$= P\left(\sup_{t \geqslant 0}(B_H(t) - ct) > x\right), \qquad (11.1)$$

where the last equality holds by the symmetry of Gaussian distributions. We consider this probability for large values of the initial capital, $x \gg 1$. In fact we simply let $x \to \infty$ to find the asymptotic behavior of the probability.

First we state the self-similarity property of the fractional Brownian motion.

Proposition 11.3. *For any positive a, the random process $\{B_H(t), t \geqslant 0\}$ and $\{a^{-H} B_H(at), t \geqslant 0\}$ have the same distribution.*

To prove this, it is sufficient to establish the equality of the means – both are zero – and the covariance functions. It is left as a simple exercise to the reader.

Now we continue with calculations (11.1), assuming $\sigma^2 = 1$ for simplicity:

$$P\left(\exists t \geqslant 0 : B_H(t) - ct > x\right) = P\left(\exists t \geqslant 0 : \frac{B_H(t)}{t^H} > \frac{x + ct}{t^H}\right)$$

$$= P\left(\exists s \geqslant 0 : \frac{B_H(sx)}{x^H s^H} > \frac{x + cxs}{x^H s^H}\right)$$

$$= P\left(\sup_{s \geqslant 0} \frac{x^{-H} B_H(sx) s^H}{s^H(1 + cs)} > x^{1-H}\right)$$

$$= P\left(\sup_{t \geqslant 0} \frac{B_H(t)}{1 + ct} > x^{1-H}\right).$$

In the second equality the change $t = xs$ is made, and in the fourth one the self-similarity property of $B_H(t)$ is used. Now let us consider the Gaussian process in the last step under the probability sign. We denote it by $Y(t)$. Its expected value is zero, and its variance is equal to

$$EY^2(t) := \sigma^2(t) = \frac{t^{2H}}{(1+ct)^2}.$$

Let us find the maximum of $\sigma(t)$ and the point where it is attained. We have,

$$\sigma'(t) = \frac{Ht^{H-1}}{1+ct} - \frac{ct^H}{(1+ct)^2} = 0,$$

so the point $t = 0$ is not the maximum point. We can safely assume then that $t > 0$. Solving the equation

$$H - \frac{ct}{1+ct} = 0,$$

we get

$$t_0 = \frac{H}{c(1-H)},$$

the unique point of maximum. To figure out the second derivative at t_0, we rewrite the first derivative as

$$\sigma'(t) = \frac{Ht^{H-1}}{1+ct} - \frac{ct^H}{(1+ct)^2} = \frac{t^{2H}}{(1+ct)^2}\left(H(1+ct)t^{-H-1} - ct^{-H}\right)$$

$$= \frac{t^{2H}}{(1+ct)^2}\left(Ht^{-H-1} - ct^{-H}(1-H)\right).$$

Differentiating,

$$-H(H+1)t^{-H-2} + cH(1-H)t^{-H-1} = Ht^{-H-2}(-H-1+c(1-H)t),$$

which is equal to $Ht_0^{-H-2}(-H-1+H) = -Ht_0^{-H-2}$ at t_0. Thus

$$\sigma''(t_0) = -\frac{Ht_0^{H-2}}{(1+ct_0)^2} = -Ht_0^{H-2}(1-H)^2.$$

So,

$$\frac{\sigma(t)}{\sigma(t_0)} = 1 - \frac{1}{2}c^2H^{-1}(1-H)^3(t-t_0)^2 + O\left((t-t_0)^3\right). \qquad (11.2)$$

Now consider the correlation function $r(s,t)$ of $B_H(t)/(1+ct)$ in the vicinity of t_0. Denote $d(t) = (\sigma(t)(1+ct))^{-1}$, and make use of the representation

$$r(s,t) = 1 - \frac{1}{2}E(d(t)B_H(t) - d(s)B_H(s))^2.$$

We have,

$$d(t)B_H(t) - d(s)B_H(s) = d(t_0)(B_H(t) - B_H(s))$$
$$+ (d(t) - d(t_0))(B_H(t) - B_H(s))$$
$$+ (d(t) - d(s))B_H(s),$$

$$Ed^2(t_0)(B_H(t) - B_H(s))^2 = d(t_0)|t - s|^{2H},$$
$$E(d(t) - d(t_0))^2(B_H(t) - B_H(s))^2 = |t - s|^{2H}O((t - t_0)^2) \text{ as } t \to t_0,$$
$$E(d(t) - d(s))^2 B_H^2(s) = O((t - s)^2) \text{ as } s, t \to t_0.$$

Therefore, if $H < 1$,

$$r(s,t) = 1 - \frac{1}{2}d^2(t_0)|t - s|^{2H}(1 + o(1)) \text{ as } s, t \to t_0.$$

Since $d(t_0) = t_0^{-H}$, we can rewrite this as

$$r(s,t) = 1 - \frac{1}{2}d(t_0)|t' - s'|^{2H}(1 + o(1)) \text{ as } s', t' \to 2^{-1/(2H)}t_0^{1/2},$$

where we set $s' = 2^{-1/(2H)}t_0^{-1/2}s$, $t' = 2^{-1/(2H)}t_0^{-1/2}t$ to fit the conditions of Theorem 10.1. So (11.2) can be rewritten as

$$\frac{\sigma(t)}{\sigma(t_0)} = 1 - \frac{1}{2}c^2H^{-1}(1 - H)^3 2^{-1/H}t_0^{-1}(t' - t_0')^2 + O\left((t' - t_0')^3\right)$$
$$= 1 - 2^{-1-1/H}c^3H^{-2}(1 - H)^2(t' - t_0')^2 + O\left((t' - t_0')^3\right).$$

It should be clear by now what we need to do. For any $T > t_0$, by Theorem 10.1 **(i)**,

$$P\left(\max_{t \in [0,T]} \frac{B_H(t)}{1 + ct} > x^{1-H}\right)$$
$$\sim \frac{\sqrt{\pi}H_{2H}c^{1/2}(1 - H)^{1/2}A^{(2-H)/2H}}{H^{1/2}B^{1/2}2^{-1/4H}} x^{(1-H)^2/H} \Psi\left(Ax^{1-H}\right), \qquad (11.3)$$

where

$$A := \sigma(t_0) = \left(\frac{H}{c(1 - H)}\right)^{-H} \frac{1}{1 - H}$$

and

$$B := -\sigma''(t_0) = \left(\frac{H}{c(1 - H)}\right)^{-H-2} H.$$

Now we want to go from the bounded interval $[0, T]$ to $[0, \infty]$. To this end, let us estimate the probability

$$P\left(\sup_{t \geq T} \frac{B_H(t)}{1 + ct} > x^{1-H}\right),$$

where T can be chosen to be a sufficiently large integer. By the entropy inequality,

$$P\left(\sup_{t \geq T} \frac{B_H(t)}{1 + ct} > x^{1-H}\right) \leq \sum_{k=T}^{\infty} P\left(\sup_{t \in [k,(k+1)]} \frac{B_H(t)}{1 + ct} > x^{1-H}\right)$$

$$\leq Cx^d \sum_{k=T}^{\infty} \exp\left(-\frac{1}{2}x^{2-2H}\frac{(1 + ck)^2}{k^{2H}}\right)$$

for some C and d. Since $\frac{(1+ck)^2}{k^{2H}} \sim c^2 k^{2-2H}$ as $k \to \infty$, passing from the sum to an integral, we get that the sum tends to zero faster then $\exp\left(-Cx^{2-2H}\right)$ with any sufficiently large C, provided T is sufficiently large. Thus the relation (11.3) holds for the infinite interval as well.

12

Excursion Probabilities: Generalizations

The Pickands double sum method developed in Lectures 9 – 11 can be readily generalized to other classes of Gaussian functions, including Gaussian random fields. Here is a typical blueprint for this. First, an "informative part" of the parameter set, the set where the trajectories reach their maximal values with high probability, is determined. In the case of stationary processes and fields all the points from the parameter set are "the same", so the whole parameter set is informative; in the case of non-stationary functions one usually chooses an infinitely small neighbourhood of the point(s) of maximum variance. This is similar to the Laplace method for evaluating asymptotics of integrals of the type

$$\int_T f(x) \exp(\lambda A(x)) dx \text{ as } \lambda \to \infty.$$

Here, we would first investigate the set of maximum points of the amplitude $A(x)$, and its behavior near these points. A small neighbourhood would then be chosen as the informative set.

A fundamental distinction between our case and this "elementary", or traditional, Laplace method is that we have to deal with functionals in infinite-dimensional spaces, so that we integrate over trajectories of random Gaussian functions in the probability space. Moreover, the amplitude function in the traditional Laplace method is typically "well behaved", i.e. has some kind of smoothness. In our case we deal with indicators of sets of trajectories that exceed a high level, so there is not much to say about their smoothness. Nevertheless, the second step in the double sum method is also similar to the classical asymptotic method. We split the set of trajectories of interest into subsets of trajectories that reach their maximum values on small subsets of the informative set. In the case of Gaussian stationary processes satisfying Pickands' condition as in Lecture 9, such subsets are intervals with lengths of order $u^{-2/\alpha}$. In the case of non-stationary processes from Lecture 10, the lengths of such small intervals vary depending on the interplay between correlations at zero and variances at their maximum points. Using small intervals is convenient as one can use the local analysis of conditional Gaussian distributions: a trajectory cannot go too far off over a small time interval. This analysis contributes the main term of the excursion probability, as the main lemma of Lecture 9 states. Finally, we need to account for those trajectories that go over a high level in several, two or more, small intervals. Because of the optimal choice of the lengths of the small intervals, the probability of the set of such trajectories is negligible. In the stationary case this step is performed by estimating the double sum in the Bonferroni inequality. In other cases we have to use different approaches that we review in this lecture.

12.1 Locally Stationary Processes with Constant Variances

Let us first consider processes with a constant mean and variance, although similar definitions can be given for more general cases.

Definition 12.1. *A Gaussian process $X(t)$ with zero mean and unit variance defined on a compact $T \subset \mathbb{R}$ is called* locally stationary *if there exists a positive continuous function C_t on T and a positive $\alpha \leqslant 2$ such that the covariance function $r(s,t)$ satisfies*

$$\lim_{t_1,t_2 \to t} \frac{1 - r(C_t t_1, C_t t_2)}{|t_2 - t_1|^\alpha} = 1$$

uniformly in $t \in T$.

Theorem 12.2. *For a locally stationary Gaussian process $X(t)$ on a compact T,*

$$P(\max_T X(t) > u) = \left(\int_T C_t^{-1} dt \right) H_\alpha u^{2/\alpha} \Psi(u)(1 + o(1))$$

as $u \to \infty$.

Proof. Let us split T into sets that are at most δ, $\delta > 0$, in diameter, $T = \cup U_i$, diam $U_i \leqslant \delta$, $i = 1, \ldots, N < \infty$. On U_i, denote

$$C_i^- = \inf_{t \in U_i} C_t, \quad C_i^+ = \sup_{t \in U_i} C_t,$$

and consider a pair of Gaussian stationary processes $X_\pm(t)$, $t \in U_i$, with zero means, unit variances and covariance functions

$$r_\pm(t) = 1 - (C_i^\pm)^{1/\alpha} |t|^\alpha + o(|t|^\alpha), \quad t \to 0,$$

respectively. By Pickands' theorem,

$$P(\max_{U_i} X_{\pm}(t) > u) = (C_i^{\pm})^{-1}|U_i|H_{\alpha}u^{2/\alpha}\Psi(u)(1 + o(1))$$

as $u \to \infty$. By Slepian's inequality,

$$\frac{1}{C_i^+}|U_i| = \lim_{u\to\infty} \frac{P(\max_{U_i} X_+(t) > u)}{H_{\alpha}u^{2/\alpha}\Psi(u)} \leqslant \liminf_{u\to\infty} \frac{P(\max_{U_i} X(t) > u)}{H_{\alpha}u^{2/\alpha}\Psi(u)}$$

$$\leqslant \limsup_{u\to\infty} \frac{P(\max_{U_i} X(t) > u)}{H_{\alpha}u^{2/\alpha}\Psi(u)} \leqslant \lim_{u\to\infty} \frac{P(\max_{U_i} X_-(t) > u)}{H_{\alpha}u^{2/\alpha}\Psi(u)} = \frac{1}{C_i^-}|U_i|.$$

For a small $\varepsilon > 0$, take δ small enough that the following holds,

$$\sum_i \frac{1}{C_i^+} \geqslant (1 - \varepsilon)\int \frac{dt}{C_t} \quad \text{and} \quad \sum_i \frac{1}{C_i^-} \leqslant (1 + \varepsilon)\int \frac{dt}{C_t}.$$

From here,

$$\limsup_{u\to\infty} \frac{P(\max_{t\in T} X(t) > u)}{H_{\alpha}u^{2/\alpha}\Psi(u)}$$

$$\leqslant \limsup_{u\to\infty} \frac{\sum_i P(\max_{t\in T} X(t) > u)}{H_{\alpha}u^{2/\alpha}\Psi(u)} \leqslant (1 + \varepsilon)\int \frac{dt}{C_t}.$$

Furthermore,

$$P(\max_{t\in T} X(t) > u)$$

$$\geqslant \sum_i P(\max_{t\in U_i} X(t) > u) - \sum_{i,j:i\neq j} P(\max_{t\in U_i} X(t) > u, \max_{t\in Uj} X(t) > u).$$

The double sum $\Sigma_2(u)$ is estimated exactly with the same scheme as in Theorem 10.1 (case $\alpha < \beta$), see (10.2). Following the arguments and inequalities there, we obtain that

$$\limsup_{u\to\infty} \frac{\Sigma_2(u)}{u^{2/\alpha}\Psi(u)} = 0.$$

Therefore, by the above inequalities,

$$\liminf_{u\to\infty} \frac{P(\max_{t\in T} X(t) > u)}{H_{\alpha}u^{2/\alpha}\Psi(u)} \geqslant (1 - \varepsilon)\int \frac{dt}{C_t}.$$

Since ε can be arbitrarily small, the theorem follows.

12.2 Several Equal Maxima of Variance

Theorem 12.3. *Let the variance of a Gaussian random process on a compact set T reach its global maximum at a finite number of points. Assume* *that in the neighborhood of any of them the assumptions **E1** and **E2** are fulfilled, perhaps with different α, β and a. Let the assumption **E3** be also fulfilled. Then the asymptotic behavior of the excursion probability for large u of the process is equal to the sum of the expressions given in (i) – (iii) of Theorem 10.1.*

The proof should be pretty evident, via an elementary application of the Bonferroni inequality. Around each of the maximum points we extract the informative sets, as in the first step of the proof in Lecture 10; then we show that the excursion probability on the rest of the parametric set is negligibly small as $u \to \infty$. Then we apply the Bonferroni inequality for the union of all informative intervals and show, using **E3**, that the corresponding double sum is also infinitely smaller because of the positive distances between the informative intervals. As the final step we apply (i) – (iii) of Theorem 10.1.

12.3 Gaussian Homogeneous Fields

A Gaussian homogeneous field on a Euclidean space is a Gaussian random function with constant mean and variance and the covariance function that depends only on the difference of its arguments. Let us now consider a question on how to generalize Pickands' condition

$$r(t) = 1 - C|t|^\alpha + o(|t|^\alpha),$$

that, as we know, is the only essential condition for the double sum method to work. For example, a product of one-dimensional covariance functions is a covariance function of a vector argument, and so we have some covariance functions of the type

$$r(\mathbf{t}) = 1 - \sum_{j=1}^{d} |t_j|^{\alpha_j} + o\left(\sum_{j=1}^{d} |t_j|^{\alpha_j}\right),$$

$\mathbf{t} = (t_1, \dots, t_d)$. One has to prove the existence of a correlation function with

$$r(\mathbf{t}) = 1 - ||\mathbf{t}||^\alpha + o(||\mathbf{t}||^\alpha).$$

Problem 12.4 (Not easy!). Prove that $r(\mathbf{t}) = \exp(-||\mathbf{t}||^\alpha)$, $\alpha \in (0,2]$, is a covariance function.

A possible generalization of Pickands' condition is as follows. Let $\boldsymbol{\alpha} = \{\alpha_1, \dots \alpha_k\}$, $k > 0$, $k \leqslant d$, be a collection of positive numbers each not higher then two, and $E = \{e_1, \dots, e_k\}$ be a collection of integers with $\sum e_i = d$. We call the pair $(E, \boldsymbol{\alpha})$ a *structure*. For a d-dimensional vector $\mathbf{t} = (t_1, \dots, t_d)$, we define the *structural module*,

$$||\mathbf{t}||_{E,\alpha} = \sum_{i=1}^{k} \left(\sum_{j=E(i-1)+1}^{E(i)} t_j^2 \right)^{\alpha_i/2},$$

where $E(i) = \sum_{j=0}^{i} e_j$, with the sum over the empty set assumed to be zero. Denote $\mathbf{t}^i = (t_{E(i-1)+1}, \ldots, t_{E(i)})$, $i = 1, \ldots, k$, then

$$||\mathbf{t}||_{E,\alpha} = \sum_{i=1}^{k} ||\mathbf{t}^i||^{\alpha_i},$$

where the summands are the usual Euclidean norms. The structure (E, α) generates the decomposition of $\mathbb{R}^d$ into the direct product of spaces $\mathbb{R}^{e_i}$,

$$\mathbb{R}^d = \bigotimes_{i=1}^{k} \mathbb{R}^{e_i},$$

so that the restriction of the structural module onto any of the subspaces is the usual Euclidean norm taken to the respective degree α_i, $i = 1, \ldots, k$. Introduce now a linear transform of $\mathbb{R}^d$,

$$g_u \mathbb{R}^d = \bigotimes_{i=1}^{k} u^{-2/\alpha_i} \mathbb{R}^{e_i}, \tag{12.1}$$

which is a superposition of the dilations $\mathbb{R}^{e_i}$ respectively, with the corresponding dilation coefficients. It should be evident that

$$||g_u \mathbf{t}||_{E,\alpha} = u^{-2} ||\mathbf{t}||_{E,\alpha}.$$

Let us consider a Gaussian field $\chi(\mathbf{t})$ on $\mathbb{R}^d$ with the mean

$$E\chi(\mathbf{t}) = -||\mathbf{t}||_{E,\alpha}$$

and the covariance function

$$\text{Cov}(\chi(\mathbf{s}), \chi(\mathbf{t})) = ||\mathbf{s}||_{E,\alpha} + ||\mathbf{t}||_{E,\alpha} - ||\mathbf{t} - \mathbf{s}||_{E,\alpha}.$$

The proof of the following lemma is similar to its one-dimensional analogue, Lemma 9.1, but uses definitions introduced above to replace the absolute value of a real number. This proof, in a more general setting, can be found in Piterbarg [1996], Lemma 6.1.

Lemma 12.5. *Let $X(\mathbf{t})$, $\mathbf{t} \in \mathbb{R}^d$, be an a.s. continuous Gaussian homogeneous field with zero mean and a covariance function that satisfies*

$$r(\mathbf{t}) = 1 - ||\mathbf{t}||_{E,\alpha} + o(||\mathbf{t}||_{E,\alpha}), \quad \mathbf{t} \to 0$$

for some structural module (E, α). Then for any compact set $T \subset \mathbb{R}^d$,

$$P(\max_{t \in g_u T} X(t) > u) = H_{E,\alpha}(T)\Psi(u)(1 + o(1))$$

as $u \to \infty$, where

$$H_{E,\alpha}(T) = E \exp(\max_{t \in T} \chi(t)) < \infty.$$

Notice that the Gaussian field $\chi(t)$ can be represented as a sum of k independent Gaussian fields $\chi(t^i)$ with means $-||t^i||^{\alpha_i}$ and covariance functions $||s^i||^{\alpha_i} + ||t^i||^{\alpha_i} - ||t^i-s^i||^{\alpha_i}$, respectively, defined on the corresponding subspaces from the direct product (12.1). Therefore, if T is a multi-dimensional rectangle with edges parallel to the coordinate axes, then

$$H_{E,\alpha}(T) = \prod_{i=1}^{k} H_{\alpha_i}(T \cap \mathbb{R}^{e_i}),$$

where $H_{\alpha_i}(T \cap \mathbb{R}^{e_i})$ is defined similarly. The following theorem is proved in Piterbarg [1996] (Lemma 7.1). The proof is similar to the proof of Pickands' theorem, Theorem 9.15.

Theorem 12.6. *Let $X(t)$, $t \in T \subset \mathbb{R}^d$, with T the closure of on open bounded set, be a Gaussian homogeneous field with zero mean and the covariance function satisfying the conditions,*

$$r(Ct) = 1 - ||t||_{E,\alpha} + o(||t||_{E,\alpha}), \quad t \to 0$$

and

$$r(t - s) < 1$$

for all s, t with $||s - t|| \in (0, \mathrm{Diam}(T)]$ for some structure (E, α) and a non-degenerate matrix C. Then

$$P(\max_{t \in T} X(t) > u) = H_{E,\alpha}|\det C|^{-1}|T|\prod_{i=1}^{k} u^{2e_i/\alpha_i}\Psi(u)(1 + o(1))$$

as $u \to \infty$, where $|T| > 0$ is the volume of T and

$$H_{E,\alpha} = \lim_{\Lambda \to \infty} \frac{H_{E,\alpha}([0,\Lambda]^d)}{\Lambda^d}, \quad 0 < H_{E,\alpha} < \infty.$$

We note that this theorem can also be generalized to Gaussian locally homogeneous fields where we still require them to have constant mean and variance. The corresponding definition is similar, as we only need to replace the continuous positive function $C(t)$ by a continuous uniformly positive-definite matrix function $C(t)$ and introduce a structural modulus (E, α). The result that generalizes Theorem 12.2 states that

$$P(\max_{t \in T} X(t) > u) = H_{E,\alpha} \int_T |\det C_t|^{-1} \, dt \prod_{i=1}^{k} u^{2e_i/\alpha_i}\Psi(u)(1 + o(1)) \quad (12.2)$$

as $u \to \infty$.

12.4 Further Generalizations

12.4.1 Fields with a Unique Point of Maximum Variance

An important question is how to generalize the assumptions **E1** and **E2** of Lecture 10 to Gaussian fields. It seems natural to define local structures to describe the behavior of the corresponding covariance function at zero and variance at the maximum point, say (E, α) and (E_1, β). If these structures are similar in the sense that $E = E_1$, the evaluations of respective asymptotic relations is analogous to the proof of Theorem 10.1, although much more complicated technically, see Piterbarg [1996], Theorem 8.1. In the case of different E and E_1 the problem becomes more complicated yet and has no complete solution still.

12.4.2 Fields on Smooth Manifolds

Let $\mathcal{T}$ be a smooth compact manifold in $\mathbb{R}^d$, and $X(\mathbf{t})$ be a Gaussian field defined on $\mathbb{R}^d$ or on a compact set of non-zero measure containing $\mathcal{T}$. One often needs to study asymptotic behavior of

$$P(\max_{\mathbf{t} \in \mathcal{T}} X(\mathbf{t}) > u)$$

as $u \to \infty$. As a rule, the idea of "summing up the infinitesimals" works here as well. We can partition $\mathcal{T}$ onto small compacts ensuring first that this partition is an atlas, that is, its every element (chart) can be smoothly mapped one-to-one to the Euclidean space of respective dimension; and second, we must be sure that one of the above results can be applied to the charts. Then we can use semi-additivity of excursion probabilities to apply the Bonferroni inequalities. We shall see an example of this later on, when we consider a Gaussian field on a Euclidean spherical cylinder.

12.4.3 Gaussian Vector Processes

Let $\mathbf{X}(t) = (X_1(t), \dots, X_d(t))$ be a Gaussian vector process, that is, a vector random function such that any finite-dimensional distribution of it is Gaussian. Let $\|\cdot\|$ be a norm in a d-dimensional vector space $\mathbb{L}^d$, with a scalar product $\langle \cdot, \cdot \rangle$. Denote by $\mathbb{M}^d$ the dual space generated by this scalar product. Then for any vector $\mathbf{x} \in \mathbb{L}^d$, by the duality property of the norm,

$$\|\mathbf{x}\| = \max_{\mathbf{l} \in S_d} (\mathbf{l}, \mathbf{x}),$$

where S_d is the unit sphere in $\mathbb{M}^d$, so that

$$P(\max_{t \in T} \|\mathbf{X}(t)\| > u) = P(\max_{(t,\mathbf{l}) \in T \times S_d} \langle \mathbf{l}, \mathbf{X}(t) \rangle > u).$$

Thus the problem of asymptotic behavior of the excursion probability for a norm of a Gaussian vector process is reduced to the problem for a Gaussian field on a manifold.

Example: χ^2-process

This example is from one of my papers Piterbarg [1994], where complete proofs could be found.

Let $(X_1(t),\ldots,X_d(t))$ be independent identically distributed Gaussian stationary processes satisfying the conditions of Pickands' theorem, Theorem 9.15. Denote $\chi^2(t) = \sum_{j=1}^{d} X_j^2(t)$. Then, denoting $\mathbf{l} = (l_1,\ldots,l_d)$,

$$P(\max_{t\in T} \chi(t) > u) = P\left(\max_{(t,\mathbf{l})\in T\times S_d} \sum_{j=1}^{d} l_j X_j(t) > u\right).$$

The variance of the Gaussian field $Y(\mathbf{l},t) = \sum_{j=1}^{d} l_j X_j(t)$ indexed on the cylinder $T \times S_d$ equals 1, its mean equals zero, the covariance function is

$$R((\mathbf{l},t),(\mathbf{l}_1,t_1)) = E(Y(\mathbf{l},t),Y(\mathbf{l}_1,t_1)) = r(t - t_1)\langle\mathbf{l},\mathbf{l}_1\rangle.$$

In order to apply our asymptotic results we want to index the field Y by points of a set from an Euclidean space with dimension d, since the dimension of the cylinder $T \times S_d$ equals d. The simplest way to do this is to substitute

$$l_1 = l_1(l_2,\ldots,l_d) = \sqrt{1 - \sum_{2}^{d} l_j^2},$$

and get the field $Y(l_1(l_2,\ldots,l_d),l_2,\ldots,l_d,t)$ indexed on the compact $T \times \{\sum_{2}^{d} l_j^2 \leqslant 1\}$ of positive measure. This field is locally homogeneous with constant variance 1. The related structure on $T \times \{\sum_{2}^{d} l_j^2 < 1\}$ is $E = (1,1,\ldots,1)$, $\boldsymbol{\alpha} = (\alpha,2,\ldots,2)$. The covariance function of Y can now be rewritten as $(v_i = l_{i+1}, i = 1,\ldots,d-1)$

$$R(t,\mathbf{v}) = 1 - |t|^\alpha - \frac{1}{2}\sum_{1}^{d-1} v_i^2 + o(|t|^\alpha + \sum_{1}^{d-1} v_i^2), \quad (t,\mathbf{v}) \to 0.$$

Applying (12.2),

$$P(\max_{t\in T} \chi(t) > u) = H_\alpha(2\pi)^{-(d-1)/2}V_{d-1}(S_{d-1})|T|u^{d-1+2/\alpha}\Psi(u)(1 + o(1)),$$

$u \to \infty$, where $V_{d-1}(S_{d-1}) = 2\pi^{d/2}/\Gamma(d/2)$ is the $(d-1)$-volume of the unit sphere, and turning to $\chi^2(t)$,

$$P(\max_{t\in T} \chi^2(t) > u)$$
$$= 2^{-d/2+1/2}H_\alpha\pi^{1/2}\Gamma(d/2)^{-1}|T|u^{d/2-1+1/\alpha}\exp(-u/2)(1 + o(1)),$$

as $u \to \infty$.

Example: Generalized χ^2-process, Piterbarg [1996]

Let now $(X_1(t),\ldots,X_d(t))$ be independent copies of a Gaussian process satisfying the conditions **E1** – **E3** of Lecture 10. Consider the random process

$$\chi_{\mathbf{b}}^2(t) = \sum_{j=1}^{d} b_j^2 X_j^2(t), \ t \in T,$$

where T contains a neighborhood of zero. Assume that

$$b_1^2 > b_2^2 \geqslant \cdots \geqslant b_d^2. \tag{12.3}$$

The argument of the previous example leads to the excursion probability for the Gaussian random field

$$Y(\mathbf{l},t) = \sum_{j=1}^{d} b_j l_j X_j(t)$$

on the same cylinder. The variance of the field

$$\sigma^2(t) = \sum_{j=1}^{d} b_j^2 l_j^2$$

attains its global maximum at two points, $(0,\pm 1,0,\ldots,0)$. Thus the multidimensional generalization of Theorem 10.1 works. However, if instead of (12.3) we had

$$b_1^2 = b_2^2 = \cdots = b_k^2 > \cdots \geqslant b_d^2$$

with $k > 1$, the maximum of $\sigma^2(t)$ would be reached on a sub-manifold (sub-cylinder) of dimension $k - 1$, and our method would no longer work.

13

Level Crossings

We start this lecture by discussing two remarkable properties of the trajectories of Gaussian processes that highlight their "real randomness". Then we count the number of crossings of a level by Gaussian trajectories.

13.1 Absence of Tangencies. Local Maxima of Equal Heights

Our intention is to prove that almost all trajectories of a Gaussian process cannot be tangent to any deterministic function. We should however first define what being tangent means in our context. Let $f(t)$, $t \in T$, be a function on a topological space T. We say that f is tangent to zero at t if $f(t) = 0$ and there exists a neighborhood $U \subset T$ of t such that f does not change its sign on U. We say furthermore that f is tangent to a function u if $f - u$ is tangent to zero. The meaning being "tangent from above" and "tangent from below" should also be clear. For a random function we, of course, need to specify the probability of being tangent, or just say "almost sure" in the right place.

Theorem 13.1. *Let $X(t)$, $t \in T$, be an a.s. continuous Gaussian function on a sequentially compact metric space T. Let $u(t)$, $t \in T$, be a continuous deterministic function. If $\mathrm{Var} X(t) > 0$ for all t, then with probability one trajectories of $X(t)$ are not tangent to $u(t)$.*

Corollary 13.2. *Under the conditions of Theorem 13.1, for any open bounded set $I \subset T$, the distribution of $\sup_{t \in I} X(t)$ is continuous.*

Proof. Let us start with the corollary. Notice that since $X(t)$ is a.s. continuous and I is bounded, $\mathrm{P}(\sup_{t \in I} X(t) < \infty) = 1$. Furthermore, by Theorem 13.1, for any u, $\mathrm{P}(\sup_{t \in I} X(t) = u) = 0$, as otherwise we would have a tangency to $u(t) = u$.

Now let us move on to the proof of the theorem itself. We denote $\mu(t) = \mathrm{E}X(t)$, $\sigma^2(t) = \mathrm{Var}X(t)$, and consider the random process

$$Y(t) = (X(t) - u(t))/\sigma(t) + \lambda, \ t \in T,$$

where λ is a constant to be chosen later. We have, $m(t) = \mathrm{E}Y(t) = [\mu(t) - u(t)]/\sigma(t) + \lambda$, $s^2(t) = \mathrm{Var}Y(t) \equiv 1$. Let I be the set from the corollary. Let us choose λ to be large enough so that $m(t) \geqslant 0$ on the closure of I. The tangencies of the trajectories of X to u from below or from above correspond to the same tangencies of the trajectories of Y to the constant λ. Therefore it is sufficient to show the absence of tangencies Y to the level λ on a ball from a separable base of T. So, let now U be such a ball, and let $T(n) = \{t_1, \ldots, t_n\} \subset U$ be a finite subset of U. Then the random variable $\max_{T(n)} Y(t)$ has the density φG_n, where φ is the standard Normal density, and G_n is a non-decreasing function. Let us prove this statement. Noting that the probability density of $Y(t)$ equals $\varphi(x - m(t))$, we have,

$$\frac{d}{dx}\mathrm{P}(\max_{T(n)} Y(t) \leqslant x)$$

$$= \sum_{j=1}^{n} \frac{1}{\sqrt{2\pi}} e^{-\frac{1}{2}(x - m(t_j))^2} \mathrm{P}(Y(t_i) \leqslant x \text{ for all } i \neq j \mid Y(t_j) = x)$$

$$=: \varphi(x) \sum_{j=1}^{n} e^{xm(t_j) - \frac{1}{2}m^2(t_j)} G_{t_j}(x).$$

All $m(t_j)$ are non-negative. Let us show that $G_{t_j}(x)$ does not decrease in x. In fact, we will show that for any Gaussian vector $(X_0, X_1, \ldots, X_n)$ with zero mean, the conditional probability $\mathrm{P}(X_1 \leqslant u, \ldots, X_n \leqslant u \mid X_0 = u)$ does not increase in u. If the covariance function of the vector is degenerate, we find a sub-vector of maximal dimension with a non-degenerate covariance vector. Let $(X_0, \ldots, X_k)$, be such a vector. Then, for some θ_{ij}, $X_i = \sum_{j=0}^{k} \theta_{ij} X_j$, $j = k+1, \ldots, n$. The conditional distribution of $(X_1, \ldots, X_k)$ given X_0 is a Gaussian multidimensional distribution with mean vector $(r_{01}u, \ldots, r_{0k}u)$, where r_{ij} are the covariances of the components X_i, X_j, and the covariance matrix is given by $r_{ij} - r_{0i}r_{0j}$, $i, j = 1, \ldots, k$. Thus $\mathrm{P}(X_1 \leqslant u, \ldots, X_n \leqslant u \mid X_0 = u)$ is an integral of the respective density over the set

$$\left\{ x_1 \leqslant u, \ldots, x_k \leqslant u, \sum_{j=0}^{k} \theta_{ij} x_j \leqslant u, \ j = k+1, \ldots, n \right\}.$$

Centering the density with the change $y_i = x_i - r_{0i}u$, we get an integral of a function which does not depend of u, over the set

$$\{y_1 \leqslant u(1 - r_{01}), \ldots, y_k \leqslant u(1 - r_{0k}),$$

$$\sum_{j=0}^{k} \theta_{ij} y_j \leqslant u(1 - \sum_{j=0}^{k} \theta_{ij} r_{0j}), \ j = k+1, \ldots, n\}.$$

All the coefficients of u are positive because of $r_{ij} \leqslant 1$ and $\sum_{j=0}^{k} \theta_{ij} r_{0j} = \sum_{j=0}^{k} \theta_{ij} EX_i X_j = EX_0 X_i \leqslant 1$. Thus this conditional probability does not decrease in u. Therefore, G_n does not decrease in x either.

Now, since the probability density of $\max_{T(n)} Y(t)$ indeed equals φG_n with non-decreasing G_n, we let n go to infinity in such a way that T_n approaches a set that is countable and dense in U. Suppose that there exists x_0, an atom of the distribution of $\sup_{t \in U} X(t)$. Then $G_{T_n}(x_0)$ is unbounded as $n \to \infty$. But this is impossible, because of

$$\left(\int_{x_0}^{\infty} \varphi(x) dx \right) G_{t_n}(x_0) \leqslant \int_{x_0}^{\infty} \varphi(x) G_{t_n}(x) dx \leqslant 1.$$

Therefore the distribution function has no atoms.

Another surprising property of Gaussian processes is that they cannot have local maxima with equal heights! This result was first proved by Jeankyung Kim and David Pollard, Kim and Pollard [1990].

Theorem 13.3. *Let $X(t)$, $t \in T$, be an a.s. continuous Gaussian process on a sequentially compact metric space T. If $\mathrm{Var}(X(s) - X(t)) > 0$ for all $s \neq t$, then, with probability one, any trajectory of X does not have two local maxima having equal heights at different points of T.*

Proof. It is sufficient to prove the theorem when T is compact. There exists a countable family $\mathcal{K}$ of closed balls such that any open set is a union of balls from this family. If a trajectory has two local maxima with equal heights at two different points of T, then there exists two disjoint balls from $\mathcal{K}$ with the same maximum of the trajectory on each of these two balls. Thus we need to prove that for any pair of two disjoint balls K_0 and K_1,

$$P(\sup_{t \in K_0} X(t) = \sup_{t \in K_1} X(t)) = 0.$$

By compactness, this problem reduces to the following: prove that for any two different points $t_0, t_1 \in T$, there exist neighborhoods $U_0 \ni t_0$ and $U_1 \ni t_1$ with

$$P(\sup_{t \in U_0} X(t) = \sup_{t \in U_1} X(t)) = 0. \tag{13.1}$$

Let $r(s,t)$ be the covariance function of X. We have from the conditions of the theorem that

$$r(t_0, t_0) - 2r(t_0, t_1) + r(t_1, t_1) > 0,$$

that is, $r(t_0, t_1)$ cannot simultaneously be equal to both $r(t_0, t_0)$ and $r(t_1, t_1)$. To be specific, let $r(t_0, t_0) \geqslant r(t_1, t_1)$, then, surely, $r(t_0, t_0) > r(t_0, t_1)$ and $r(t_0, t_0) > 0$. By continuity of trajectories, the function

$$h(t) = r(t, t_0)/r(t_0, t_0)$$

is continuous. Consider a Gaussian function $Y(t) = X(t) - h(t)X(t_0)$. Computing its covariances, or using the properties of conditional Gaussian distributions, we obtain that this function does not depend of $X(t_0)$. Since $h(t_0) > h(t_1)$, one can find neighborhoods U_0 and U_1 such that

$$\inf_{t \in U_0} h(t) =: \beta_0 > \beta_1 := \sup_{t \in U_1} h(t).$$

Let us fix a continuous trajectory of Y. The function

$$\Gamma_0(x) = \sup_{t \in U_0}(Y(t) + h(t)x)$$

is a supremum of the set of linear functions with the slopes that are at least β_0, therefore $\Gamma_0(x)$ is concave and its right derivative is equal to at least β_0 everywhere. By the same argument, a similar function Γ_1 on U_1 exists with the right derivative equal to at least β_1 everywhere. Thus the equality $\Gamma_0(x) = \Gamma_1(x)$ can occur at at most one point x. Since the random variable $X(t_0)$ is non-degenerate and does not depend of Y, we obtain that

$$P(\sup_{t \in U_0} X(t) = \sup_{t \in U_1} X(t) \mid Y) = P(\Gamma_0(X(t_0)) = \Gamma_1(X(t_0)) \mid Y) = 0.$$

Averaging this inequality over all trajectories of Y, we obtain (13.1).

13.2 Number of Level Crossings

Let us first consider zero crossings by a continuous deterministic function. We say that a continuous function $f(t)$ crosses zero (has a zero crossing) at the point t_0 if any neighborhood of the point contains two points t_1 and t_2 with $f(t_1)f(t_2) < 0$. To be specific, let f be given on $[0,1]$. Let us denote by C the number of zero crossings by this function on the interval. Let $f(k2^{-n}) \neq 0$ for all $k = 0,1,\ldots,2^n$ and $n = 1,2,\ldots$. If $f(t_1)f(t_2) < 0$ for some points $t_1 < t_2$, then f intersects zero at least once on (t_1, t_2). Let us introduce some auxiliary variables,

$$U_{nk} = \begin{cases} 1, & \text{if } f((k-1)2^{-n})f(k2^{-n}) < 0, \\ 0, & \text{otherwise}, \end{cases}$$

$k = 1,2,\ldots,2^n$, $n = 1,2,\ldots$, and

$$Z_n = \sum_{k=1}^{2^n} U_{nk}. \tag{13.2}$$

As Z_n is a non-decreasing sequence, let us denote the limit $Z = \lim_n Z_n$. Since, as has been already mentioned, $Z_n \leqslant C$, we see that $Z \leqslant C$ as well.

Lemma 13.4 (N. D. Ylvisaker). *For the function f defined above we have that $Z = C$, where both sides can be infinite.*

Proof. If $C < \infty$, the zero crossing points are separated, hence for sufficiently large n, $C = Z_n$, and so $Z = C$. If $Z < \infty$ then, beginning with some n we have that $Z_n = Z$, and f cannot change its sign on the intervals with $U_{nk} = 0$. Thus all intersections can be counted by the number of intervals with $U_{nk} = 1$. Taking the limit $n \to \infty$, we find that there exist points $0 = t_0 < t_1 < \cdots < t_{Z+1} = 1$ such that f does not change its sign on each of the intervals $[t_i, t_{i+1}]$, $i = 0, 1, \ldots, Z$. Thus $C < \infty$.

Notice that if $C < \infty$, the up-crossings alternate with the down-crossings, because between these points f does not change its sign.

Let us now go back to Gaussian processes. We will need the following lemma.

Lemma 13.5. *Let $\rho(t, s)$ be a correlation function and $A(x)$ be a function such that $A(x) = (1 + o(1))\sqrt{2(1 - x)}$ as $x \to 1$. Then for all t, $h^{-1}A(\rho(t, t + h))$ tends to a finite limit as $h \to 0$ if and only if the second derivative $\rho_{ts}''(t, t)$ exists. In this case, $h^{-1}A(\rho(t, t + h)) \to \sqrt{\rho_{ts}''(t, t)}$.*

Proof. Let us consider a random process X with zero mean, unit variance and the covariance function $\rho(t, s)$. If the limit exists, then

$$\frac{2(1 - \rho(t, t + h))}{h^2} = (1 + o(1))h^{-2}A^2(\rho(t, t + h)).$$

Therefore,

$$\lim_{h \to 0} E\left(\frac{X(t + h) - X(t)}{h}\right)^2 = \lim_{h \to 0} \frac{2(1 - \rho(t, t + h))}{h^2} = \lim_{h \to 0} h^{-2}A^2(\rho(t, t + h)),$$

that is, $X(t)$ is differentiable in the mean square sense at t. Therefore

$$h^{-1}A(\rho(t, t + h)) \to \sqrt{\rho_{ts}''(t, t)} \text{ as } h \to 0.$$

Conversely, if the derivative $\rho_{ts}''(t, t)$ exists, X is differentiable in the mean square sense at t, so

$$h^{-2}A^2(\rho(t, t + h)) = 2h^{-2}(1 - \rho(t, t + h))(1 + o(1)) \to \rho_{ts}''(t, t).$$

If the variance $\sigma^2(t)$ of a Gaussian process $X(t)$ is positive on $[0, 1]$, the trajectories of X are not equal to zero with probability one at all points $k2^{-n}$, and so one can apply Lemma 13.4, so that Z_n does not decrease with probability one and tends a.s. to C. Let us compute the expected value of C,

$$EU_{nk} = P(X((k-1)2^{-n})X(k2^{-n}) < 0)$$
$$= P(X((k-1)2^{-n}) < 0, \, X(k2^{-n}) > 0)$$
$$+ P(X((k-1)2^{-n}) > 0, \, X(k2^{-n}) < 0).$$

The probabilities do not change after dividing $X(t)$ by $\sigma(t)$, therefore it is convenient to assume immediately that the variance is identically equal to one. The mean has also been changed but is still a continuous function as σ is positive and continuous. Denote by $\rho(t,s)$, the correlation function of X. We can also assume that $\rho := \rho((k-1)2^{-n}, k2^{-n}) < 1$, otherwise the probabilities above are equal to zero. Denote $m_1 := EX((k-1)2^{-n})$, $m_2 := EX(k2^{-n})$. Since the Gaussian density satisfies the heat equation (1.6) and $\rho > -1$ we have that

$$\frac{\partial EU_{nk}}{\partial \rho} = -2\varphi_{X((k-1)2^{-n}), \, X(k2^{-n})}(0,0)$$

$$= -\frac{2}{2\pi\sqrt{1-\rho^2}} \exp\left(-\frac{m_1^2 - 2\rho m_1 m_2 + m_2^2}{2(1-\rho^2)}\right).$$

Therefore, since $EU_{nk} = 0$ for $\rho = 1$,

$$EU_{nk} = \frac{1}{\pi} \int_\rho^1 \frac{1}{\sqrt{1-h^2}} \exp\left(-\frac{m_1^2 - 2hm_1 m_2 + m_2^2}{2(1-h^2)}\right) dh. \qquad (13.3)$$

In the particular case $m(t) = EX(t) \equiv 0$,

$$EU_{nk} = \frac{1}{\pi} \int_\rho^1 \frac{dh}{\sqrt{1-h^2}} = \frac{1}{\pi} \arccos \rho. \qquad (13.4)$$

Notice that $A(x) = \arccos x$ satisfies the conditions of Lemma 13.5. If $m(t) \equiv m$,

$$EU_{nk} = \frac{1}{\pi} \int_\rho^1 \frac{1}{\sqrt{1-h^2}} \exp\left(-\frac{m^2}{1+h}\right) dh. \qquad (13.5)$$

The function

$$A(\rho) := e^{m^2/2} \int_\rho^1 \frac{1}{\sqrt{1-h^2}} \exp\left(-\frac{m^2}{1+h}\right) dh$$

also satisfies the conditions of Lemma 13.5. Indeed, integrating by parts and using the fact that $\arccos x$ is a decreasing function, we get that

$$e^{-m^2/2} A(\rho) = -\int_\rho^1 \exp\left(-\frac{m^2}{1+h}\right) d\arccos h = \arccos \rho \exp\left(-\frac{m^2}{1+\rho}\right)$$

$$- \int_\rho^1 \arccos h \exp\left(-\frac{m^2}{1+h}\right) \frac{m^2}{(1+h)^2} dh$$

$$= \arccos \rho \exp\left(-\frac{m^2}{1+\rho}\right) (1 + O(1-\rho))$$

$$= e^{-m^2/2} \arccos \rho (1 + O(1-\rho)),$$

as $\rho \to 1$. From (13.2) it follows that for a Gaussian process with the unit variance and constant mean,

$$EZ_n = \sum_{k=0}^{2^n} EU_{nk} = \frac{1}{\pi} \sum_{k=0}^{2^n} 2^{-n} e^{-m^2/2} (2^n \arccos \rho((k-1)2^{-n}, k2^{-n}))(1 + o(1))$$

$$= \frac{1}{\pi} e^{-m^2/2} \int_0^1 \sqrt{\rho_{ts}''(t,t)} dt(1 + o(1)), \text{ as } n \to \infty. \qquad (13.6)$$

From here, using also Lemmas 13.4, 13.5 and (13.1), we get the following result.

Theorem 13.6 (Ylvisaker's theorem). *Let $X(t)$, $t \in [0,T]$, be an a.s. continuous Gaussian process. Let $EX(t) \equiv m$ and $\operatorname{Var}X(t) \equiv \sigma^2 > 0$. Then with probability one its trajectories are not tangent to zero and for the number $N(0,T)$ of zero crossings by X,*

$$EN(0,T) = \frac{1}{\sigma^2 \pi} \exp\left(-\frac{m^2}{2\sigma^2}\right) \int_0^T \sqrt{r_{ts}''(t,t)} dt,$$

where r is the covariance function of X. This relation is still valid if one of its sides is equal to infinity.

In order to prove this, all that remains is to divide X by σ and use the additivity of $EN(0,T)$ to go from $[0,1]$ to $[0,T]$.

Let us now consider the general case. Assume that the functions $m(t) = EX(t)$ and $\sigma^2(t) = \operatorname{Var}X(t)$ are continuously differentiable on $[0,T]$, and that $\sigma^2(t) > 0$ for all $t \in [0,T]$. Denote $m = m((k-1)2^{-n})$, $m' = m'((k-1)2^{-n})$, then $m(k2^{-n}) = m + 2^{-n}m' + o(2^{-n})$ as $n \to \infty$. Simple calculations of (13.3) lead to

$$EU_{nk} = \frac{1}{\pi} \int_\rho^1 \frac{1}{\sqrt{1-h^2}} \exp\left(-\frac{m^2 + mm'2^{-n} + o(2^{-n})}{1+h}\right) dh$$

$$= \frac{1}{\pi} \int_\rho^1 \frac{1}{\sqrt{1-h^2}} \exp\left(-\frac{m^2}{1+h}\right)(1 + O(2^{-n})) dh.$$

Again, integrating by parts,

$$EU_{nk} = \frac{1}{\pi} e^{-m^2/2} \arccos \rho(1 + O(1-\rho))(1 + O(2^{-n})).$$

From here, by the same arguments as before, we get the following theorem.

Theorem 13.7. *Let $X(t)$, $t \in [0,T]$, be an a.s. continuous Gaussian process. Let $m(t) = EX(t)$ and $\sigma^2(t) = \operatorname{Var}X(t)$ be continuously differentiable on $[0,T]$. Assume also that $\sigma(t) > 0$ for all $t \in [0,T]$. Then, with probability one, its trajectories are not tangent to zero, and we have the following for the number $N(0,T)$ of zero crossings by the trajectories of X,*

$$EN(0,T) = \frac{1}{\pi} \int_0^T \sqrt{\rho''_{ts}(t,t)} \exp\left(-\frac{m^2(t)}{2\sigma^2(t)}\right) dt,$$

where ρ is the correlation function of X. This still holds even if one of the sides is infinity.

Suppose that $EN(0,T) < \infty$. Then $N < \infty$ a.s. Therefore, the up-crossings alternate with the down-crossings. It follows then that the expected number of up- and down-crossings are given by

$$EN^+(0,T) = EN^-(0,T) = \frac{1}{2}EN(0,T).$$

Non-Gaussian processes. Moments of the Number of Crossings

In this lecture we ask the same questions as in Lecture 13 but for non-Gaussian random processes. We see what conditions we need for the absence of tangencies and the absence of local maxima of the same height. We also derive formulas for the average number of crossings, and for higher-order moments of the number of crossings.

14.1 Bulinskaya's Theorem

Let us prove the following lemma.

Lemma 14.1. *Let* $X(t)$, $t \in [0,1]$, *be a random process with a.s. right-continuous trajectories. Denote by* $\omega_X(t)$ *the modulus of continuity,*

$$\omega_X(t) = \sup_{s,s',|s-s'| \leqslant t} |X(s) - X(s')|, \quad 0 \leqslant t \leqslant 1.$$

Then for any $\varepsilon > 0$ *one can find a deterministic function* $\omega_\varepsilon(t) \downarrow 0$ *as* $t \downarrow 0$, *with*

$$P(\omega_X(t) < \omega_\varepsilon(t), \ 0 < t \leqslant 1) > 1 - \varepsilon.$$

Proof. It follows from a.s. continuity that for any $c > 0$

$$\lim_{t \to 0} P(\omega_X(t) < c) = 1.$$

We construct the required function starting from an arbitrary sequence of positive numbers converging to zero, $c_1 > c_2 > \cdots \to 0$. For any n there exists t_n with

$$P(\omega_X(t_n) < c_n) > 1 - \varepsilon 2^{-n}.$$

Then, since the modulus of continuity does not decrease,

$$P(\omega_X(t) < c_n, \ 0 < t \leqslant t_n) > 1 - \varepsilon 2^{-n}.$$

This inequality continues to be true if we replace t_n with a smaller (positive) number, so that we can make sure that $t_n \downarrow 0$ as $n \uparrow \infty$. Let us define $\omega_\varepsilon(t)$ to be a piecewise constant function with jumps at the points t_n and values c_n at those points. Indeed,

$$1 - \mathrm{P}(\omega_X(t) < c_n, \ 0 < t \leqslant t_n, \ n = 1, 2, \ldots)$$

$$\leqslant \sum_{n=1}^{\infty} \mathrm{P}(\omega_X(t_n) \geqslant c_n) \leqslant \sum_{n=1}^{\infty} \varepsilon 2^{-n} = \varepsilon.$$

Theorem 14.2 (E. V. Bulinskaya's theorem). *Let $X(t), t \in [0,1]$, be an a.s. continuously differentiable random process (i.e. the set of continuously differentiable trajectories has probability one). Let for any t the probability density $f_t(x)$ of $X(t)$ exist and be bounded uniformly in t. Then the probability that there exists a point t such that $X(t) = X'(t) = 0$ is equal to zero. In particular, for any fixed level, the trajectories are not tangent to it with probability one. Moreover, the number of times t such that $X(t) = 0$, $t \in [0,1]$, is finite with probability one.*

Proof. Denote by $A_{h,n,k}$ the set of functions $x(t) \in C^1[0,1]$ (the space of continuously differentiable functions on $[0,1]$) whose derivatives equal zero at at least one point, say τ_x, from $[(k-1)/n, k/n]$, $1 \leqslant k \leqslant n$, such that $|x(\tau_x)| \leqslant h$. We shall use the same notation for the set of trajectories of X belonging to $A_{h,n,k}$. Let $x(\cdot) \in A_{h,n,k}$, then

$$x(k/n) = x(\tau_x) + (k/n - \tau_x)x'(\tau_x + \theta(k/n - \tau_x)), \quad 0 \leqslant \theta \leqslant 1.$$

Consequently,

$$|x(k/n)| \leqslant h + n^{-1}\omega_{x'}(n^{-1}), \tag{14.1}$$

where $\omega_{x'}$ is the modulus of continuity of the derivative x'. For any $\omega(t) \downarrow 0$ as $t \downarrow 0$, let B_ω be a family of functions $x(t)$ from $C^1[0,1]$ such that $\omega_{x'}(t) \leqslant \omega(t)$ for all $t \in [0,1]$. Then, by Lemma 14.1, for any $\varepsilon > 0$ there exists a function $\omega_\varepsilon(t) \downarrow 0$ as $t \downarrow 0$ such that $\mathrm{P}(B_{\omega_\varepsilon}) > 1 - \varepsilon/2$. For the event

$$A_h = \{X(t) : \ X'(t_x) = 0 \text{ at at least one point } t_x \text{ and } |X(t_x)| \leqslant h\},$$

we have,

$$A_h = \bigcup_{k=1}^{n} A_{h,n,k},$$

$$\mathrm{P}(A_h) \leqslant \sum_{k=1}^{n} \mathrm{P}(A_{h,n,k}B_{\omega_\varepsilon}) + \mathrm{P}(B_{\omega_\varepsilon}^c), \tag{14.2}$$

where the notation D^c is used for the complement of D. By (14.1), the first term on the right-hand side of (14.2) is at most

$$nP(|X(k/n)| \leqslant h + n^{-1}\omega_\varepsilon(n^{-1})) \leqslant 2cn(h + n^{-1}\omega_\varepsilon(n^{-1})),$$

where c is the constant that bounds the density $f_t(x)$. Since $\omega_\varepsilon(t) \to 0$ as $t \to 0$, one can choose first n_0, and then h_0 so that the last expression is at most $\varepsilon/2$. And since $P(B^c_{\omega_\varepsilon}) \leqslant \varepsilon/2$, we have that $P(A_{h_0}) \leqslant \varepsilon$. Furthermore, the event

$$A_0 = \{X(t): \ X'(t_x) = 0, \ X(t_x) = 0 \text{ at at least one point } t_x \in [0,1]\}$$

is contained in every A_h, therefore, since ε is arbitrary, $P(A_0) = 0$, which proves the first assertion of the theorem.

Let us now prove that the number of zeros of X is finite. First let us show that, if a trajectory belongs to $A^c_{h_0}$, it has at most finite number of zeros. Assume the converse, then there exists an infinite sequence of zeros $t_1, t_2, \ldots$ from $[0,1]$. Let t_0 be its limit point. Since X is a.s. continuous, $X(t_0) = 0$. The derivative of the trajectory must have a zero between any two zeros, but since it belongs to $A^c_{h_0}$, $|X(t_i)| > h_0$, and therefore t_0 is a limit point of points with $|X(t_i)| > h_0$. This contradicts the continuity assumption. Thus any trajectory having infinitely many zeros belongs to A_{h_0}. So, since $P(A_{h_0}) \leqslant \varepsilon$ and ε is arbitrary, the probability of infinitely many zeros equals zero.

Notice that by considering the process $X(t) - u(t)$, with a continuously differentiable u, the statement of the theorem is also applicable to tangencies and crossings of any continuously differentiable function.

14.2 No Local Maxima of the Same Height

Let us now consider a non-Gaussian analogue to Theorem 13.3.

Theorem 14.3. *Let $X(t)$, $t \in [0,T]$, be a.s. differentiable and its derivative X' be such that*

$$w_{X'}(\delta) = o(\delta^{1/3}) \text{ in probability as } \delta \to 0.$$

Assume that for any $s,t \in [0,T]$, $s < t$, the vector $(X'_s, X'_t, X_t - X_s)$ has a probability density $p_{s,t}$ such that for any $d > 0$,

$$K(d) := \sup_{s,t\in[0,T], |t-s|\geqslant d, (x,y,z)\in\mathbb{R}^3} p_{s,t}(x,y,z) < \infty.$$

Then

$$P\{\exists s,t \in [0,T] : \ s < t, X'(s) = X'(t) = 0, X(s) = X(t)\} = 0. \tag{14.3}$$

Proof. Assume that $T = 1$, as we can always consider $X_{t/T}$. Define the events

$$A_{k,l,n} := \left\{ \exists s \in \left[\frac{k-1}{n}, \frac{k}{n} \right], t \in \left[\frac{l-1}{n}, \frac{l}{n} \right] : X'(s) = X'(t) = 0, X(s) = X(t) \right\}.$$

By the mean value theorem, for any $s \in [(k-1)/n, k/n]$ there exists $\theta \in [0,1]$, perhaps random, with

$$|X(k/n) - X(s)| = |k/n - s| \cdot |X'(s + \theta(k/n - s)) - X'(s)| \leqslant \frac{1}{n} w_{X'}(1/n). \quad (14.4)$$

Moreover, if $X'(s) = 0$, then

$$|X'(k/n)| = |X'(k/n) - X'(s)| \leqslant w_{X'}(1/n). \quad (14.5)$$

From the relations $X(s) = X(t)$, (14.4) and a similar estimate for $|X(l/n) - X(t)|$, we get that

$$|X(k/n) - X(l/n)| \leqslant \frac{2}{n} w_{X'}(1/n). \quad (14.6)$$

By the theorem assumptions, for any $\varepsilon > 0$ and all sufficiently large n, the relation $w_{X'}(1/n) \leqslant \varepsilon n^{-1/3}$ is valid with probability exceeding $1 - \varepsilon$. Thus by (14.4) – (14.6) and boundness of the density, for any $\delta > 0$ and all $1 \leqslant k < l \leqslant n$ with $l - k + 1 \geqslant \delta n$, one can write down that for any $\varepsilon > 0$ and all sufficiently large n,

$$
\begin{aligned}
&\mathrm{P}(A_{k,l,n}) \\
&\leqslant \mathrm{P}\left((X'(k/n), X'(l/n), X(l/n) - X(k/n)) \in [-w_{X'}(1/n), w_{X'}(1/n)]^2 \right. \\
&\qquad\qquad\qquad\qquad\qquad \left. \times [-2n^{-1} w_{X'}(1/n), 2n^{-1} w_{X'}(1/n)] \right) \\
&\leqslant \varepsilon + \mathrm{P}\left((X'(k/n), X'(l/n), X(l/n) - X(k/n)) \in [-\varepsilon n^{-1/3}, \varepsilon n^{-1/3}]^2 \right. \\
&\qquad\qquad\qquad\qquad\qquad\qquad \left. \times [-2\varepsilon n^{-4/3}, 2\varepsilon n^{-4/3}] \right) \\
&\leqslant \varepsilon + 16K(\delta - 1/n)\varepsilon^3 n^{-2},
\end{aligned}
$$

therefore,

$$\max_{1 \leqslant k \leqslant k + \delta n - 1 \leqslant l \leqslant n} \mathrm{P}(A_{k,l,n}) = o(n^{-2}). \quad (14.7)$$

Now the theorem easily follows:

$$
\begin{aligned}
&\mathrm{P}\left(\exists s, t \in [0,T] \ : \ s < t, X'(s) = X'(t) = 0, X(s) = X(t) \right) \\
&= \lim_{\delta \downarrow 0} \mathrm{P}\left(\exists s, t \in [0,T] \ : \ s + \delta < t, X'(s) = X'(t) = 0, X(s) = X(t) \right) \\
&\leqslant \lim_{\delta \downarrow 0} \lim_{n \to \infty} \sum_{1 \leqslant k \leqslant k + \delta n - 1 \leqslant l \leqslant n} \mathrm{P}(A_{k,l,n}) \\
&= 0.
\end{aligned}
$$

14.3 Moments of the Number of Crossings

Theorem 14.4 (Rice's formula). *Let $X(t)$, $t \in [0,T]$, be an a.s. continuous random process. Assume that there exists a finite set $T_0 \subset [0,T]$ such that the process $X(t)$ is differentiable in square mean on $[0,T]\backslash T_0$. Assume then that for any $s,t \in [0,T]$, $s < t$, there exists a probability density $p_{s,t}(x,y)$ of the vector*

$$\left(X(t), \frac{X(t) - X(s)}{t - s} \right),$$

and for any $t \in [0,T]\backslash T_0$ and almost all x,y there exits a finite limit

$$\lim_{s \to t} p_{s,t}(x,y) = p_{t,t}(x,y).$$

Assume furthermore that

$$\sup_x \int_{-\infty}^{\infty} |y| \sup_{s,t \in [0,T], s<t} p_{s,t}(x,y) dy < \infty.$$

Then the expected number of up-crossings $N_u^+([0,T])$ and down-crossings $N_u^-([0,T])$ of a level u by trajectories of $X(t)$ are equal to

$$EN_u^+([0,T]) = \int_0^T \int_0^{\infty} y p_{t,t}(u,y) dy dt$$

and

$$EN_u^-([0,T]) = \int_0^T \int_{-\infty}^0 |y| p_{t,t}(u,y) dy dt,$$

where $p_{t,t}$ is the probability density of $(X(t), X'(t))$ in the case $X'(t)$ exists, and $p_{t,t} = 0$ otherwise.

Proof. We shall derive the expression for the up-crossings, with the other one following by replacing X with $-X$ and u with $-u$. Since for any t, $X(t)$ has a probability density, then with probability one $X(k2^{-n}) \neq u$ for all k, n, and one can apply Lemma 13.4. Let us consider the sum

$$Z_n = \sum_{k=1}^{2^n-1} U_{nk}, \tag{14.8}$$

with

$$U_{nk} = \begin{cases} 1, & \text{if } X((k-1)2^{-n}) < u, \ X(k2^{-n}) > u, \\ 0, & \text{otherwise}, \end{cases}$$

$k = 1,2,\dots,2^n$, $n = 1,2,\dots$. First we note that (compare to Lecture 13)

$$P(X((k-1)2^{-n}) \leqslant u, \ X(k2^{-n}) > u) = \int_0^{\infty} dy \int_{u-2^{-n}y}^{u} dx p_{(k-1)2^{-n},k2^{-n}}(x,y).$$

Next,

$$\lim_{n\to\infty} 2^n \int_{u-2^{-n}y}^u dx p_{(k-1)2^{-n},k2^{-n}}(x,y) = y p_{tt}(u,y),$$

where n and k tend to infinity with $k2^{-n} \to t$. By the assumptions, the function under the integral is dominated by an integrable one, therefore,

$$EN_u^+([0,1]) = \lim_{n\to\infty} EZ_n = \lim_{n\to\infty} \sum_{k=1}^{2^n} \int_0^\infty dy \int_{u-2^{-n}|y|}^u dx p_{(k-1)2^{-n},k2^{-n}}(x,y)$$

$$= \int_0^1 \int_0^\infty y p_{t,t}(u,y) dy dt.$$

As we already mentioned, the passage from $[0,1]$ to $[0,T]$ is evident.

Notice that for $N_u([0,T])$, the number of all crossings of u,

$$EN_u([0,T]) = \int_0^T \int_{-\infty}^\infty |y| p_{t,t}(u,y) dy dt.$$

Notice also that under the assumptions of Theorem 14.4, $EN_u([0,T]) < \infty$, that is $N_u([0,T]) < \infty$ a.s.

A similar argument also works for deriving the expression for the higher-order moments of the number of crossings, using the same sum (14.8). Consider the number of up-crossings, again using $T = 1$. We have,

$$EZ_n^2 = \sum_{i,j=0}^{2^n-1} EU_{ni}U_{nj}$$

$$= \sum_{i,j=0,\ i\neq j}^{2^n-1} EU_{ni}U_{nj} + \sum_{i=0}^{2^n-1} EU_{ni}^2.$$

Since $U_{ni}^2 = U_{ni}$,

$$EZ_n(Z_n - 1) = \sum_{i,j=0,\ i\neq j}^{2^n-1} EU_{ni}U_{nj}.$$

Consider the events A_ε, $\varepsilon > 0$,

$$A_\varepsilon = \left\{ \begin{array}{c} \text{the distance between any two up-crossings} \\ \text{on } [0,1] \text{ is no less than } \varepsilon \end{array} \right\}.$$

By the monotone convergence,

$$\lim_{n\to\infty} EZ_n(Z_n - 1)\mathbf{I}_{A_\varepsilon} = EN_u(N_u - 1)\mathbf{I}_{A_\varepsilon}.$$

Denote by $p_{s_1,s_2,t_1,t_2}(x_1,x_2,y_1,y_2)$, the probability density of

$$\left(X(s_1),X(s_2),\frac{X(t_1)-X(s_1)}{t_1-s_1},\frac{X(t_2)-X(s_2)}{t_2-s_2}\right)$$

and assume for any $\varepsilon > 0$,

$$\sup_{x_1,x_2}\int_{-\infty}^{\infty}\int_{-\infty}^{\infty}|y_1 y_2|\sup_{\substack{s_1,s_2,t_1,t_2\in[0,T]\\s_1<t_1,s_2<t_2,s_2-t_1>\varepsilon}}p_{s_1,s_2,t_1,t_2}(x_1,x_2,y_1,y_2)dy_1\,dy_2 < \infty.$$

(14.9)

Assume also that for almost all t_1, t_2 the limit

$$\lim_{s_1\to t_1,s_2\to t_2}p_{s_1,s_2,t_1,t_2}(x_1,x_2,y_1,y_2) =: p_{t_1,t_2}(x_1,x_2,y_1,y_2)$$

(14.10)

exists. Similar to the proof of Theorem 14.4, for all sufficiently large n and all i,j with $|i2^{-n}-j2^{-n}| > \varepsilon$, we get that

$$EU_{ni}U_{nj} = \int_0^\infty dy_1\int_0^\infty dy_2\int_{u-2^{-n}y_1}^u dx_1\int_{u-2^{-n}y_2}^u dx_2 p_{i,j,n}(x_1,x_2,y_1,y_2),$$

where

$$p_{i,j,n}(x_1,x_2,y_1,y_2) = p_{i2^{-n},j2^{-n},(i+1)2^{-n},(j+1)2^{-n}}(x_1,x_2,y_1,y_2).$$

Again, with the same argument as earlier, for almost all $t_1,t_2,|t_1-t_2| \geq \varepsilon$,

$$\lim_{n\to\infty}2^{2n}\int_{u-2^{-n}y_1}^u\int_{u-2^{-n}y_2}^u p_{i,j,n}(x_1,x_2,y_1,y_2)dx_1\,dx_2 = y_1 y_2 p_{t_1,t_2}(u,u,y_1,y_2),$$

where $n,i,j \to \infty$ in such a way, that $i2^{-n} \to t_1$ and $j2^{-n} \to t_2$. Therefore

$$\lim_{n\to\infty}EZ_n(Z_n-1)\mathbf{I}_{A_\varepsilon}$$

$$= \int_{t_1,t_2\in[0,1],|t_1-t_2|\geq\varepsilon}\int_0^\infty\int_0^\infty y_1 y_2 p_{t_1,t_2}(u,u,y_1,y_2)dy_1\,dy_2 dt_1\,dt_2.$$

Since $N_u([0,T]) < \infty$ a.s. and $\mathbf{I}_{A_\varepsilon}\uparrow 1$ as $\varepsilon\downarrow 0$, using the monotone convergence we get the following theorem.

Theorem 14.5. Let a random process $X(t)$, $t\in[0,T]$ satisfy the assumptions of Theorem 14.4. Let (14.9) and (14.10) be fulfilled. Then

$$EN_u^+([0,T])(N_u^+([0,T])-1)$$

$$= \int_0^T\int_0^T\int_0^\infty\int_0^\infty y_1 y_2 p_{t_1,t_2}(u,u,y_1,y_2)dy_1\,dy_2 dt_1\,dt_2.$$

By changing signs of the process and the level, one can get a similar formula for the second factorial moments of the number of down-crossings and for the number of all crossings. We now formulate one more theorem concerning the joint second moments of the number of crossings of two random processes.

Theorem 14.6. *Let $X_1(t)$ and $X_2(t)$, $t \in [0,T]$, be random processes satisfying the conditions of Theorem 14.4. Assume that for all t_1, t_2, s_1, s_2, $t_1 > s_1, t_2 > s_2$ the probability density $q_{t_1,t_2,s_1,s_2}(x_1,x_2,y_1,y_2)$ of*

$$\left(X_1(t_1), X_2(t_2), \frac{X_1(t_1) - X_1(s_1)}{t_1 - s_1}, \frac{X_2(t_2) - X_2(s_2)}{t_2 - s_2} \right)$$

exists and satisfies (14.9) and (14.10). Then for $N_{1,u}^+([0,T])$ and $N_{2,u}^+([0,T])$ of the up-crossings of u by the processes $X_1(t)$ and $X_2(t)$, respectively,

$$E N_{1,u}^+([0,T]) N_{2,u}^+([0,T])$$
$$= \int_0^T \int_0^T \int_0^\infty \int_0^\infty y_1 y_2 q_{t_1,t_2,t_1,t_2}(u,u,y_1,y_2) dy_1 dy_2 dt_1 dt_2,$$

where q_{t_1,t_2,s_1,s_2} is the probability density of $(X_1(t_1), X_2(t_2), X_1'(s_1), X_2'(s_2))$.

14.3.1 Remark on the Moments of the Number of Crossings for Gaussian Processes

We did not consider second moments of the number of crossings in Lecture 13. Here we present two obvious corollaries to Theorems 14.5 and 14.6 for Gaussian processes.

Corollary 14.7. *Let for a Gaussian, differentiable in square mean, random process $X(t)$, $t \in [0,T]$ and any $t_1, t_2 \in [0,T]$, $t_1 \neq t_2$, the covariance matrix of the vector $(X(t_1), X(t_2), X'(t_1), X'(t_2))$ be non-degenerate. Then the statement of Theorem 14.5 holds.*

Corollary 14.8. *Let $X_1(t)$ and $X_2(t)$, $t \in [0,T]$, be two Gaussian, differentiable in square mean, processes. Let for any $t_1, t_2 \in [0,T]$, $t_1 \neq t_2,$, the covariance matrix of the vector $\left(X_1(t_1), X_2(t_2), X_1'(t_1), X_2'(t_2) \right)$ be non-degenerate. Then the statement of Theorem 14.6 holds.*

In conclusion we point out that we did not consider the question whether the moments are actually finite. For the Gaussian processes, and only for them, the exact answer exists (Geman [1972]):

$$E N_u^+([0,T])^2 < \infty \iff \exists \delta > 0 : \int_0^\delta t^{-1}(r''(t) - r''(0)) dt < \infty.$$

There are also some generalizations for Gaussian fields, see Elizarov [1985]. Moreover, one can find the expressions for higher moments in Azaîs and Wschebor [2009].

15

Probabilities of High Excursions. The Method of Moments

In this lecture we study another method for estimating probabilities of high excursions, a method that was probably the first to be discovered. It was developed by S. O. Rice, Rice [1944], in 1944, who based his formula on Theorem 14.4. The intuition behind the method is clear. If the trajectories of a Gaussian process are smooth enough, an excursion of a trajectory above a high level on a finite time interval can happen only once. In other words, the number of up-crossings of a high level can only equal zero or one, with a very high probability. Therefore the probability of trajectories exceeding a high level can be approximated by the mean number of up-crossings. This reasonable but somewhat flawed argument requires rigorous justification, of course. In this lecture we provide one for Gaussian processes and show that

$$P(\max_{t \in [0,T]} X(t) > u) \sim EN_u^+(0,T) \qquad (15.1)$$

as $u \to \infty$. Moreover, we show that this method not only allows us to estimate the accuracy of the approximation, but also to obtain an asymptotic decomposition of the probability in (15.1).

We first consider processes that are not necessarily Gaussian but satisfy the conditions of Lecture 14. We assume, like in Bulinskaya's theorem, that a random process $X(t)$, $t \in [0,T]$, is a.s. continuously differentiable, and its one-dimensional probability densities are bounded uniformly in t. Assume also that for any level u, the variance of the number of u-crossings by $X(t)$ is finite,

$$EN_u^2(0,T) < \infty. \qquad (15.2)$$

We have,

$$P(\max_{t \in [0,T]} X(t) > u) = P(X(0) > u) + P(X(0) \leqslant u, \max_{t \in [0,T]} X(t) > u). \qquad (15.3)$$

Since there are no tangencies, the last probability can be written as

$$P(X(0) \leqslant u, \max_{t \in [0,T]} X(t) > u)$$

$$= P(X(0) \leqslant u, N^+ = 1) + P(X(0) \leqslant u, N^+ \geqslant 2)$$
$$= P(N^+ = 1) - P(N^+ = 1, X(0) > u)$$
$$+ P(X(0) \leqslant u, N^+ \geqslant 2), \tag{15.4}$$

where $N^+ = N_u^+(0,T)$ is, as before, the number of up-crossings of the level u by the trajectories of $X(t)$. The last summand can be written as

$$P(N^+ = 1) = EN^+ - \sum_{k=2}^{\infty} k p_k,$$

where $p_k = P(N^+ = k)$. Furthermore,

$$P(X(0) \leqslant u, N^+ \geqslant 2) = \sum_{k=2}^{\infty} p_k - P(X(0) \geqslant u, N^+ \geqslant 2),$$

and summing up,

$$P(X(0) \leqslant u, \max_{t \in [0,T]} X(t) > u)$$

$$= EN^+ - \sum_{k=2}^{\infty} (k-1) p_k - P(X(0) \geqslant u, N^+ \geqslant 2).$$

Thus

$$P(X(0) \leqslant u, \max_{t \in [0,T]} X(t) > u) \leqslant EN^+,$$

and by (15.3),

$$P(\max_{t \in [0,T]} X(t) > u) \leqslant EN^+ + P(X(0) > u). \tag{15.5}$$

In order to bound the probability from below we write

$$P(N^+ = 1, X(0) > u) =$$
$$P(X(0) > u, X(T) > u, N^+ = 1)$$
$$+ P(X(0) > u, X(T) \leqslant u, N^+ = 1)$$
$$\leqslant P(X(0) > u, X(T) > u) + P(N_u^-(0,T) \geqslant 2)$$
$$\leqslant P(X(0) > u, X(T) > u) + \frac{1}{2} EN^-(N^- - 1),$$

where $N^- = N_u^-(0,T)$ is the number of respective down-crossings. Together with (15.4) it gives,

$$P(\max_{t\in[0,T]} X(t) > u) \tag{15.6}$$

$$\geqslant EN^+ + P(X(0) > u) - P(X(0) > u, X(T) > u)$$

$$- \frac{1}{2}EN^-(N^- - 1).$$

Thus the relation (15.1) holds if

$$P(X(0) > u) + EN^-(N^- - 1) + P(X(0) > u, X(T) > u) = o(EN^+)$$

as $u \to \infty$.

Let us state the result that we just obtained.

Lemma 15.1. *Assume that a.s. continuously differentiable random process $X(t)$, $t \in [0,T]$, satisfies either Bulinskaya's theorem conditions if it is non-Gaussian, or Theorem 13.7 conditions if it is Gaussian. Assume (15.2). Then*

$$0 \geqslant P(\max_{t\in[0,T]} X(t) > u) - EN_u^+(0,T) - P(X(0) > u)$$

$$\geqslant -\frac{1}{2}EN_u^-(0,T)(N_u^-(0,T) - 1) - P(X(0) > u, X(T) > u).$$

Interestingly, the excursion probability that is an "integral" in an infinite-dimensional space can be approximated by integrals of probability densities in one and two time points.

From now on we should deal with Gaussian stationary processes only. Let $X(t)$, $t \in [0,T]$, be a Gaussian stationary process with zero mean, unit variance and the covariance function $r(t)$ such that

$$r(t) = 1 - \frac{1}{2}t^2 + Ct^4 + o(t^4) \tag{15.7}$$

as $t \to 0$. By Corollary 14.7,

$$EN_u^-(0,T)(N_u^-(0,T) - 1)$$

$$= T \int_0^T t \int_{-\infty}^0 \int_{-\infty}^0 |y_1 y_2| \varphi_t(u,u,y_1,y_2) dy_1 dy_2 dt. \tag{15.8}$$

Let us estimate the integral

$$I(t) = \int_{-\infty}^0 \int_{-\infty}^0 |y_1 y_2| \varphi_t(u,u,y_1,y_2) dy_1 dy_2,$$

where, as earlier, $\varphi_t(x_1,x_2,y_1,y_2)$ is the probability density of

$$(X(0),X(t),X'(0),X'(t)).$$

Here is the idea of the estimate. Given $X(0) = X(t) = u$ with a high level u, it is likely that $X'(0) > 0$, especially for small t, whereas the upper limits of $I(t)$ are set to zero. Thus, the main area of focus of the behavior of the integration function should be near $y_1 = 0$.

Some additional difficulties could arise while analyzing the multiplier $|y_1 y_2|$. To avoid these difficulties let us take $B > 0$, the value that we choose later, and write,

$$I(t) = \int_{-B}^{0} \int_{-B}^{0} + \int \int_{\min(y_1,y_2) \leqslant -B} |y_1 y_2| \varphi_t(u,u,y_1,y_2) dy_1 dy_2$$

$$\leqslant B^2 \int_{-B}^{0} \int_{-B}^{0} \varphi_t(u,u,y_1,y_2) dy_1 dy_2$$

$$+ \int \int_{\min(y_1,y_2) \leqslant -B} |y_1 y_2| \varphi_t(u,u,y_1,y_2) dy_1 dy_2$$

$$\leqslant B^2 \int_{-B}^{0} \varphi_{1t}(u,u,y_1) dy_1 + B \int_{-\infty}^{-B} |y_1| \varphi_{1t}(u,u,y_1) dy_1$$

$$+ B \int_{-\infty}^{-B} |y_2| \varphi_{2t}(u,u,y_2) dy_2, \tag{15.9}$$

where $\varphi_{1t}(x_1,x_2,y)$ is the density of $(X(0),X(t),X'(0))$, and $\varphi_{2t}(u,u,y)$ is the density of $(X(0),X(t),X'(t))$. Now let us pass to the conditional densities given $X(0) = X(t) = u$. Continuing (15.9),

$$I(t) \leqslant B\varphi_t(u,u) \left(B \int_{-B}^{0} \varphi_{1t}(y_1 \mid u,u) dy_1 + \int_{-\infty}^{-B} |y_1| \varphi_{1t}(y_1 \mid u,u) dy_1 \right.$$

$$\left. + \int_{-\infty}^{-B} |y_2| \varphi_{2t}(y_2 \mid u,u) dy_2 \right). \tag{15.10}$$

Here

$$\varphi_t(u,u) = \frac{1}{2\pi\sqrt{1 - r^2(t)}} \exp\left(-\frac{u^2}{1 + r(t)}\right)$$

is the density of $(X(0),X(t))$ at (u,u), with $\varphi_{it}(y \mid u,u)$, $i = 1,2$, being, respectively, the conditional densities of $X'(0)$ and $X'(t)$ given $X(0) = X(t) = u$. The corresponding conditional expected values are

$$m(t)u = \mathrm{E}(X'(0) \mid X(0) = X(t) = u) = \frac{-r'(t)u}{1 + r(t)},$$

$$-m(t)u = \mathrm{E}(X'(t) \mid X(0) = X(t) = u) = \frac{r'(t)u}{1 + r(t)},$$

and the conditional variances coincide,

$$\sigma^2(t) = \text{Var}(X'(0) \mid X(0) = X(t) = u)$$

$$= \text{Var}(X'(t) \mid X(0) = X(t) = u) = \frac{2 - 2r^2(t) - r'(t)^2}{1 - r^2(t)}.$$

First let us consider small t, $t \leqslant \delta$, where the value δ will be chosen later. Consider the first integral in the brackets on the right-hand side of (15.10). We denote it by I_1. It is equal to

$$P(X'(0) \leqslant 0 \mid X(0) = X(t) = u) = \Phi\left(-\frac{um(t)}{\sigma(t)}\right),$$

where Φ is the standard Gaussian distribution function. By Taylor's formula,

$$\frac{m(t)}{\sigma(t)} \sim (6C - 1)^{-1/2}, \quad t \to 0,$$

and, moreover, $6C - 1 > 0$, because

$$4!C - 4 = \text{E}(X(t) + \sqrt{2}X''(t))^2 > 0.$$

Here the expected value cannot be zero since, by assumptions on the existence of multidimensional densities, $X(t)$ cannot be a sinusoid. Now let us take δ small enough so that

$$\frac{m(t)}{\sigma(t)} \geqslant k(6C - 1)^{-1/2} =: \delta_1 > 0 \tag{15.11}$$

for all $t \in [0,\delta]$, where the positive $k < 1$ can be used to optimize the estimate above for the second factorial moment. Thus for $t \in [0,\delta]$,

$$I_1(t) \leqslant \frac{1}{\sqrt{2\pi\delta_1 u}} \exp(-\delta_1^2 u^2/2).$$

Let us bound by one the integral for remaining t,

$$I_1(t) \leqslant 1, \quad t > \delta.$$

The second integral, $I_2(t)$, in the brackets on the right-hand side of (15.10) (φ is the standard Normal density) can be written as

$$I_2(t) = -\int_{-\infty}^{-B} y_1 \varphi\left(\frac{y - um(t)}{\sigma(t)}\right) dy$$

$$= -\int_{-\infty}^{-B-um(t)/\sigma(t)} \sigma(t)(\sigma(t)y + um(t)) \, \varphi(y) \, dy.$$

Denote

$$M(T) = \max_{t\in[0,T]} |m(t)/\sigma(t)|, \quad M_m(T) = \max_{t\in[0,T]} |m(t)|, \quad M_\sigma(T) = \max_{t\in[0,T]} |\sigma(t)|.$$

Obviously, $M(T)$, $M_m(T)$, $M_\sigma(T) < \infty$. Now take $B = 2M(T)u$, use $\varphi'(x) = -x\varphi(x)$ and get for all t, that

$$I_2(t) \leqslant M_\sigma^2(T) \int_{-\infty}^{-M(T)u} y\varphi(y)\, dy + M_\sigma(T)M_m(T)u\Phi(-M(T)u)$$

$$\leqslant \left(\frac{1}{\sqrt{2\pi}}M_\sigma^2(T) + \frac{M_\sigma(T)M_m(T)}{\sqrt{2\pi}M(T)}\right)e^{-M^2(T)u^2/2} =: \mathcal{M}e^{-M^2(T)u^2/2}.$$

$$(15.12)$$

Comparing the above expressions for the conditional expectations and variances one can see that the last inequality also holds for $I_3(t)$, the third integral on the right-hand side of (15.10).

All that remains now is to estimate $\varphi_t(u, u)$. For $t \in [0, \delta]$,

$$\varphi_t(u, u) \leqslant \frac{1}{2\pi\sqrt{1 - r^2(t)}}e^{-u^2/2},$$

for $t \geqslant \delta$, $r(t) \leqslant \max_{t \geqslant \delta} r(t) =: \tilde{r}(\delta) < 1$, therefore

$$\varphi_t(u, u) \leqslant \frac{1}{2\pi\sqrt{1 - \tilde{r}^2(\delta)}}\exp\left(-\frac{u^2}{1 + \tilde{r}(\delta)}\right).$$

Bringing together all the estimates we obtain that

$$I(t) \leqslant \pi^{-1}M(T)u\left(\frac{\mathbf{I}_{\tau \leqslant \delta}}{\sqrt{1 - r^2(t)}}e^{-u^2/2} + \frac{\mathbf{I}_{\tau \geqslant \delta}}{\sqrt{1 - \tilde{r}^2(\delta)}}\exp\left(-\frac{u^2}{1 + \tilde{r}(\delta)}\right)\right)$$

$$\times \left(M(T)u\mathbf{I}_{\tau \geqslant \delta} + \frac{M(T)\mathbf{I}_{\tau \leqslant \delta}}{\sqrt{2\pi}\delta_1}\exp(-\delta_1^2 u^2/2) + (M(T)u + 1)\mathcal{M}e^{-M^2(T)u^2/2}\right)$$

$$\leqslant \pi^{-1}M(T)u\left(\frac{\exp\left(-(1 + \delta_1^2)u^2/2\right)}{\sqrt{2\pi}\delta_1\sqrt{1 - r^2(t)}} + \frac{M(T)u\exp\left(-u^2/(1 + \tilde{r}(\delta))\right)\mathbf{I}_{\tau \geqslant \delta}}{\sqrt{1 - \tilde{r}^2(\delta)}}\right.$$

$$\left. +(M(T)u + 1)\mathcal{M}\frac{\exp\left(-(1 + M^2(T))u^2/2\right)}{\sqrt{2\pi}\delta_1\sqrt{1 - r^2(t)}}\right)$$

$$\leqslant \frac{Cu^2}{\sqrt{1 - r^2(t)}}\exp\left(-\frac{(1 + \Delta)u^2}{2}\right)$$

for

$$\Delta = \min\left(\delta_1^2, \frac{1 - \tilde{r}(\delta)}{1 + \tilde{r}(\delta)}, M^2(T)\right) \qquad (15.13)$$

and some positive constant C (the maximum of all the coefficients in front of the exponents). Integrating in (15.8), we obtain that

$$EN_u^-(0, T)(N_u^-(0, T) - 1) \leqslant CT^2 u^2 \exp\left(-\frac{(1 + \Delta_1)u^2}{2}\right),$$

which, together with Lemma 15.1 and the formula for the expected value of the up-crossings from Lecture 13, gives us the following result.

Theorem 15.2. *Assume that the covariance function of a Gaussian stationary process $X(t)$ with zero mean satisfies (15.7). Let the conditions of Corollary 14.7 be fulfilled. Then*

$$P(\max_{t\in[0,T]} X(t) > u) = \frac{T}{2\pi}e^{-u^2/2} + 1 - \Phi(u) + R(u),$$

where

$$0 < R(u) \leqslant \frac{1}{2}(C+2)T^2 u^2 \exp\left(-\frac{(1+\Delta)u^2}{2}\right),$$

where $\Delta > 0$ is defined in (15.13).

In order to conclude the proof, we just notice that

$$P(X(0) > u, X(T) > u) \leqslant P(X(0) + X(T) > 2u)$$
$$\leqslant \frac{1}{\sqrt{2\pi}\sqrt{2+2r(T)}}e^{-u^2/(1+r(T))},$$

and that

$$\frac{1}{1+r(T)} \geqslant \frac{1}{1+\tilde{r}(\delta)}.$$

For smooth Gaussian stationary processes, Rice's method gives rather more information on the behavior of high excursion probabilities than Pickands' method. In fact, here we have an asymptotic expansion of this probability up to exponentially smaller infinitesimals. Similar expansions can be obtained by Rice's method for non-stationary Gaussian processes as well, and even for Gaussian fields, but estimates become much more complicated, and the main lemma is, of course, different.

To finish off this lecture we note that as a side effect of our discourse we computed Pickands' constant H_α for $\alpha = 2$,

$$H_2 = \frac{1}{\sqrt{\pi}}.$$

Poisson Limit Theorem for Large Values of Stationary Sequences

In this and the future lectures we come back to the Poisson limit theorem for the number of large values of Gaussian stationary processes. In this lecture we prove this theorem for Gaussian stationary sequences, considering the large values as a stochastic point process and giving necessary and sufficient conditions for the process to be asymptotically Poisson. The monograph Kallenberg [1986] is strongly recommended for a deeper study of the theory of point processes.

Let $X(k), k \in \mathbb{Z}$, be a Gaussian stationary sequence with zero mean, unit variance and the covariance function $r(k)$. In this lecture we study the limit behavior of the random set of points

$$\{k \in \mathbb{Z} : X(k) > u\}$$

when $u \to \infty$. If there were a bounded set here instead of $\mathbb{Z}$, we could immediately say that this random set converges to the empty set with probability one, since $P(X(t) > u) \to 0$ as $u \to \infty$. To obtain non-trivial results we shall usually consider expanding subsets of $\mathbb{Z}$ as u increases. We start our investigation by defining a point process.

Let $\mathbb{X}$ be a locally compact second-countable Hausdorff space, see Kallenberg [1986] or Willard [1970]. Let us fix a metric on $\mathbb{X}$ (the space we consider admits a metric), and denote by $\mathcal{B} = \mathcal{B}(\mathbb{X})$ the σ-ring of bounded Borel subsets of $\mathbb{X}$. Let N be a set of all non-negative integer-valued locally-bounded measures on $\mathcal{B}$. Denote by $\mathcal{N}$ the minimal σ-algebra of subsets of N such that for any $B \in \mathcal{B}$, the map

$$\mu \to \mu B$$

from $(N, \mathcal{N})$ to $\mathbb{Z}_+ = \{0,1,2,\ldots\}$ is measurable (we assume that $\mathbb{Z}_+$ is equipped with the σ-ring of all of its bounded subsets.) A random *point process* on $\mathcal{B}$ is a measurable map from the canonical probability space $(\Omega, \mathcal{F}, P)$ to $(N, \mathcal{N})$. That is, any $\omega \in \Omega$ corresponds to a non-negative integer-valued measure $\xi(B,\omega)$, $B \in \mathcal{B}$, with the usual measurability properties of random variables. Thus it is natural to call the distribution of a random point process a (random) measure $P\xi^{-1}$ on $(N, \mathcal{N})$. A weak convergence of a sequence of point processes $\xi_n \overset{w}{\to} \xi$ is the weak convergence of their distributions. Now we are ready to introduce random point processes of interest to us here. Recall that

$$P(X(1) > u) \sim \Psi(u) := \frac{\exp(-u^2/2)}{\sqrt{2\pi}u}, \quad u \to \infty.$$

Consider point processes on $\mathcal{B}(\mathbb{R})$, the σ-ring of all bounded Borel sets in $\mathbb{R}$,

$$\xi_u(B) = \xi_u(B,\omega) = \sum_{k \in B/\Psi(u)} I_{\{X(k)>u\}}(\omega), \tag{16.1}$$

$\omega \in \Omega$, $u > 0$, $B \in \mathcal{B}(\mathbb{R})$, where the notation $B/\Psi(u)$ denotes k for which $k\Psi(u) \in B$, and I is the indicator function. Since B is bounded, the sum above is well-defined and finite. Let us denote $m_u(B) = E\xi_u(B)$ and $d_u(B) = \text{Var}\xi_u(B)$, both finite. It is easy to see that m is a measure proportional to the Lebesgue measure.

Now let us look at how one should prove weak convergence of point processes. We need an analogue to the Prokhorov theorem and its corollaries.

Definition 16.1. *A ring $\mathcal{L} \subset \mathcal{B}$ is called a DC-ring (DC stands for "dissecting covering") if for any bounded closed set $B \in \mathcal{B}$ and arbitrary $\varepsilon > 0$ there exists a finite covering of B by sets $L \in \mathcal{L}$ such that $\text{diam}B \leqslant \varepsilon$. A semi-ring with the same properties is called a DC-semi-ring.*

Let us fix a DC-semi-ring $\mathcal{U} \subset \mathcal{B}$ and a DC-ring $\mathcal{L} \subset \mathcal{B}$. A random point process is called *simple* if its atoms have measure 1 with probability 1, that is, if $\xi(\{t\},\omega) > 0$ then $\xi(\{t\},\omega) = 1$ with probability 1. We denote $\mathcal{B}_\xi = \{B \in \mathcal{B} : \xi(\partial B) = 0 \text{ a.s.}\}$, ∂B is a boundary of B.

Theorem 16.2 (O. Kallenberg, Kallenberg [1986]). *Let ξ_n be a sequence of arbitrary random point processes on $\mathcal{B}(\mathbb{X})$, ξ be a simple random point process on $\mathbb{X}$. Also assume that there exists a DC-semi-ring $\mathcal{U} \subset \mathcal{B}_\xi$ and a DC-ring $\mathcal{L} \subset \mathcal{B}_\xi$. Then $\xi_n \overset{w}{\to} \xi$ if*

1. $\forall L \in \mathcal{L}$, $\lim_{n\to\infty} P(\xi_n(L) = 0) = P(\xi(L) = 0)$;
2. $\forall U \in \mathcal{U}$, $\lim\sup_{n\to\infty} E\xi_n(U) \leqslant E\xi(U) < \infty$.

Let us denote by $\varphi(x,y;r)$ the two-dimensional Gaussian density with zero means, unit variances and the covariance r. Let us introduce two functionals on the sequence $\{r(k), k \geqslant 0\}$:

$$R(r,L) = \sum_{k \in L/\Psi(u), r(k)>0} (N-k)r(k) \int_0^1 \varphi(u,u; hr(k))dh,$$

$$R_\varepsilon(r,L) = \sum_{k \in L/\Psi(u):\ r(k) \geqslant \varepsilon u^{-2}} (N-k)r(k) \int_{\varepsilon/u^2 r(k)}^1 \varphi(u,u; hr(k))dh,$$

where $N = \text{Card}(B/\Psi(u))$. Here and in what follows we assume that if $h = |r| = 1$, then $\varphi = 0$, even though the two-dimensional density does not exists in this case. As the integral of $1/\sqrt{1-h^2}$ converges, this is consistent with the definition.

A random point process $\pi(B,\omega)$ where the distribution $\mathrm{P}\pi^{-1}(\cdot)$ is Poisson is called a *Poisson point process* with the intensity measure $\lambda(B)$, $B \in \mathcal{B}$, if for any B,

$$P(\pi(B) = k) = \frac{\lambda(B)^k}{k!}e^{-\lambda(B)}, \; k = 0,1,2,\ldots .$$

If a Poisson point process is given on an Euclidean space and its intensity measure $\lambda(B) = \mathrm{E}\pi(B)$ is absolutely continuous with respect to the Lebesgue measure, then this Poisson point process is simple (the reader should prove this). If the intensity measure is proportional to Lebesgue measure, $\lambda(B) = \lambda l(B)$, the process π is called a *stationary (homogeneous) Poisson process* with intensity λ. Notice that a ring generated by semi-intervals of finite length is a DC-ring on $\mathcal{B}(\mathbb{R})$. Let us denote it by $\mathcal{L}$.

Theorem 16.3. *Let $X(k)$ be a Gaussian stationary sequence with zero mean and unit variance. Let π be a stationary Poisson point process on $\mathbb{R}$ with the intensity λ. Then the following are equivalent,*

(1). $\xi_u \overset{w}{\to} \pi$, $u \to \infty$;
(2). $\forall L \in \mathcal{L}, \; \lim_{u \to \infty} m_u(L) = \lambda l(L) = \lim_{u \to \infty} d_u(L)$;
(3). $\forall L \in \mathcal{L}, \; \lim_{u \to \infty} m_u(L) = \lambda l(L)$ *and* $\lim_{u \to \infty} R(r,L) = 0$;
(4). $\forall L \in \mathcal{L}, \; \lim_{u \to \infty} m_u(L) = \lambda l(L)$ *and* $\forall \varepsilon > 0 \; \lim_{u \to \infty} R_\varepsilon(r,L)) = 0$.

Proof. Without loss of generality we assume that $\lambda = 1$.

1. $(2) \Rightarrow (4)$. For the sake of clarity we prove this implication for $L = [1,0)$, as the general case is treated in the same way. We have,

$$\text{Var} \sum_{k \in L/\Psi(u)} I_{\{X(k)>u\}}(\omega) = \sum_{k \in L/\Psi(u)} \text{Var}I_{\{X(k)>u\}}(\omega)$$
$$+ \sum_{k,l \in L/\Psi(u), k \neq l} \text{Cov}\left(I_{\{X(k)>u\}}, I_{\{X(l)>u\}}\right)$$
$$= \sum_{k \in L/\Psi(u)} P(X(k) > u) - \sum_{k \in L/\Psi(u)} P(X(k) > u)^2$$
$$+ \sum_{k,l \in L/\Psi(u), k \neq l} \text{Cov}\left(I_{\{X(k)>u\}}, I_{\{X(l)>u\}}\right)$$
$$\to 1$$

as $u \to \infty$. The first sum on the right tends to $l(L) = 1$, and the second one evidently tends to zero by the dominated convergence theorem. Therefore, necessarily,

$$\sum_{k,l \in L/\Psi(u), k \neq l} \text{Cov}\left(I_{\{X(k)>u\}}, I_{\{X(l)>u\}}\right) \to 0.$$

Furthermore, like in the proof of the comparison theorem (Theorem 2.6) denoting $r = r(k - l)$, we obtain that

$$
\begin{aligned}
\mathrm{Cov}\left(I_{\{X(k)>u\}}, I_{\{X(l)>u\}}\right) \\
= P(X(k) > u, X(l) > u) - P(X(k) > u)P(X(l) > u) \\
= \int_0^1 \frac{d}{dh} P(X_h(k) > u, X_h(l) > u)dh \\
= r \int_0^1 \varphi(u, u; hr)dh.
\end{aligned}
\tag{16.2}
$$

For the Gaussian two-dimensional density with $h < 1$ if $r = 1$, and all h otherwise,

$$
\varphi(u, u; hr) = \frac{1}{2\pi\sqrt{1 - h^2 r^2}} \exp\left(-\frac{u^2}{1 + hr}\right) \leqslant \frac{1}{2\pi\sqrt{1 - h^2 r^2}} e^{-u^2}
$$

as $r \leqslant 0$. Therefore

$$
\sum_{k,l\in L/\Psi(u), k\neq l} \mathrm{Cov}\left(I_{\{X(k)>u\}}, I_{\{X(l)>u\}}\right)
$$

$$
= \sum_{k,l\in L/\Psi(u), k\neq l, r(k-l)>0} r \int_0^1 \varphi dh + \sum_{r<0} r \int_0^1 \varphi dh
$$

$$
\geqslant \sum_{r\geqslant 0} r \int_0^1 \varphi dh + \frac{1}{2\pi} \sum_{r<0} r \int_0^1 \frac{e^{-u^2}|r|dh}{\sqrt{1 - h^2 r^2}}.
\tag{16.3}
$$

The second summand (see (16.2)) is

$$
\sum_{k,l\in L/\Psi(u), k\neq l, r(k-l)<0} \mathrm{Cov}\left(I_{\{X_h(k)>0\}}, I_{\{X_h(l)>0\}}\right) e^{-u^2/2} e^{-u^2/2}.
$$

Since the covariance function $r(k)$ is non-negative-definite, the sum over all r is non-negative, therefore the sum over negative r can be estimated from below by the sum

$$
- \sum_{k,l\in L/\Psi(u), k\neq l, r(k-l)>0} \mathrm{Cov}\left(I_{\{X_h(k)>0\}}, I_{\{X_h(l)>0\}}\right) e^{-u^2/2} e^{-u^2/2},
$$

which is equal to

$$
- \sum_{k,l=0, k\neq l, r(l-k)\geqslant 0}^{1/\Psi(u)} \frac{r}{2\pi} \int_0^1 \frac{e^{-u^3}dh}{\sqrt{1 - h^2 r^2}} - \sum_{k=0}^{1/\Psi(u)} \mathrm{Cov} I_{\{X_h(k)>0\}} e^{-u^2}.
$$

The variance does not depend on k and the second sum tends to zero. Let us denote it by α_u. Continuing (16.3),

$$\ldots \geq \sum_{r>0} r \int_0^1 \varphi dh - \sum_{r\geq 0} \frac{r}{2\pi} \int_0^1 \frac{e^{-u^3} dh}{\sqrt{1-h^2 r^2}} + \alpha_u \qquad (16.4)$$

$$= \sum_{r>0} r \int_0^1 \varphi \cdot \left(1 - \exp\left(-u^2 + \frac{u^2}{1+rh}\right)\right) dh + \alpha_u$$

$$= \sum_{r>0} r \int_0^1 \varphi \cdot \left(1 - \exp\left(-\frac{u^2 rh}{1+rh}\right)\right) dh + \alpha_u$$

$$\geq \sum_{r>0} r \int_{\varepsilon/u^2 r}^1 \varphi \cdot \left(1 - \exp\left(-\frac{u^2 rh}{1+rh}\right)\right) dh + \alpha_u$$

$$\geq \alpha_u + \left(1 - e^{-\varepsilon/2}\right) \sum_{ru^2 \geq \varepsilon} r \int_0^1 \varphi dh.$$

The last sum is just $R_\varepsilon(r,[0,1))$ and it converges to zero as $u \to \infty$ by the conditions. This proves (2)$\Rightarrow$(4).

2. (3)$\Rightarrow$(4). Obvious.

3. (4)$\Rightarrow$(3). Keeping in mind that $r \geq 0$, we write,

$$\sum_{r<\varepsilon u^{-2}} (N-k)r \int_0^1 \frac{1}{2\pi\sqrt{1-h^2 r^2}} \exp\left(-\frac{1}{2} \frac{2u^2 - 2u^2 hr}{1-h^2 r^2}\right) dh$$

$$\leq \sum_{|r|<\varepsilon u^{-2}} \frac{N-k}{2\pi u^2} u^2 re^{-u^2} \int_0^1 \frac{1}{\sqrt{1-h^2 r^2}} \exp\left(\frac{hru^2}{1+hr}\right) dh$$

$$\leq \sum_{|r|<\varepsilon u^{-2}} \frac{N-k}{2\pi u^2} e^{-u^2} u^2 r \int_0^1 \frac{1}{\sqrt{1-h^2 r^2}} \exp\left(\frac{h\varepsilon}{1-h\varepsilon u^{-2}}\right) dh$$

$$\leq \frac{2e^{1/2}}{\sqrt{3}} \left[\varepsilon \Psi^2(u) \sum_{k=1}^{1/\Psi(u)} (N-k) \right],$$

for $\varepsilon < 1/2$ and $u \geq 1$. The fraction in front of the square brackets bounds the last integral, the expression in the square brackets converges to ε since $\Psi(u) \sim P(X(k) > u)$. Furthermore,

$$\sum_{r\geq \varepsilon u^{-2}} (N-k)r \int_0^1 \varphi dh = \sum_{r\geq \varepsilon u^{-2}} (N-k)r \left(\int_0^{\varepsilon/ru^2} \varphi dh + \int_{\varepsilon/ru^2}^1 \varphi dh\right)$$

$$= \sum_{r\geq \varepsilon u^{-2}} (N-k)r \int_0^{\varepsilon/ru^2} \varphi dh + \sum_{r\geq \varepsilon u^{-2}} (N-k)r \int_{\varepsilon/ru^2}^1 \varphi dh.$$

The second sum is $R_\varepsilon(r(\cdot),[0,1))$, and it converges to zero by the conditions of the theorem. For the first sum, where $r > 0$,

$$\frac{r}{2\pi}\int_0^{\varepsilon/ru^2}\frac{1}{2\pi\sqrt{1-h^2r^2}}\exp\left(-\frac{1}{2}\frac{2u^2-2u^2hr}{1-h^2r^2}\right)dh$$

$$=\frac{1}{2\pi u^2}e^{-u^2}ru^2\int_0^{\varepsilon/ru^2}\frac{1}{\sqrt{1-h^2r^2}}\exp\left(\frac{u^2rh}{1+hr}\right)dh$$

$$=\frac{1}{2\pi u^2}e^{-u^2}\int_0^{\varepsilon}\frac{1}{\sqrt{1-s^2u^{-4}}}\exp\left(\frac{s}{1+su^{-2}}\right)ds,$$

with $s = rhu^2$. The integral is uniformly bounded for all sufficiently large u by a function $\gamma(\varepsilon)$ that tends to zero as $\varepsilon \to 0$. Therefore, the sum is bounded by $2\gamma(\varepsilon)$. Since ε is arbitrary, (4)$\Rightarrow$(3) follows.

4. (3)$\Rightarrow$(1). Let us show that for any $L \in \mathcal{L}$, $P(\xi_u(L) = 0)$ converges to $\exp(-l(L))$ as $u \to \infty$, so that the required weak convergence follows from Kallenberg's theorem. Let $Z(k)$, $k \in \mathbb{Z}$, be a sequence of independent standard Gaussian variables that does not depend on the sequence $X(k)$. Denote by $\zeta_u(B, \omega)$ the corresponding to $Z(k)$ point process defined similarly to (16.1). Let us find a bound for the difference

$$\Delta P = P(\xi_u(L) = 0) - P(\zeta_u(L) = 0).$$

Notice that $P(\xi_u(L = 0) = P(\max_{k \in L/\Psi(u)]} X(k) \leqslant u)$, and the same holds true for ζ_u. By Theorem 2.6, denoting $X_h(k) = \sqrt{h}X(k) + \sqrt{1-h}Z(k)$,

$$\Delta P = \sum_{k,l \in L/\Psi(u)\ k>l,} r(k-l)$$

$$\times \int_0^1 \varphi(u,u,r_h(k-l))P(X_h(j) \leqslant u, j \neq k,l \mid X_h(k) = X_h(l) = u)dh$$

$$=\sum_{r>0} r \int_0^1 \varphi(u,u,r_h(k-l))P(X_h(j) \leqslant u, j \neq k,l \mid X_h(k) = X_h(l) = u)dh$$

$$-\sum_{r<0} |r| \int_0^1 \varphi(u,u,r_h(k-l))P(X_h(j) \leqslant u, j \neq k,l \mid X_h(k) = X_h(l) = u)dh.$$

The first sum on the right-hand side does not exceed $R(r,L)$. The second sum does not exceed $\sum_{r<0} |r| \int_0^1 \varphi dh$. Per (16.3), (16.4) we have,

$$\sum_r r \int_0^1 \varphi dh \geqslant R_\varepsilon(r,L)(1 - e^{-\varepsilon/2}) + \alpha_u,$$

and since

$$\sum_r r \int_0^1 \varphi dh = \sum_{r>0} r \int_0^1 \varphi dh - \sum_{r<0} |r| \int_0^1 \varphi dh,$$

using the already proven equivalence (3)$\Leftrightarrow$(4), we have that

$$\sum_{r<0} |r| \int_0^1 \varphi dh \leqslant \sum_{r>0} r \int_0^1 \varphi dh - R_\varepsilon(r,L)(1 - e^{-\varepsilon/2}) - \alpha_u \to 0$$

as $u \to \infty$. Therefore

$$|\Delta P| \leqslant \sum_r r^+ \int_0^1 \varphi dh + \sum_r r^- \int_0^1 \varphi dh \to 0.$$

Now we can apply the classic Poisson limit theorem to $\zeta_u(L)$, keeping in mind that

$$\lim_{u\to\infty} E\zeta_u(L) = \lim_{u\to\infty} m_u(L) = \lambda l(L).$$

Thus $P(\zeta_u([0,1]) = 0) \to e^{-\lambda}$ as $u \to \infty$. The condition 2) of Kallenberg's Theorem obviously follows from the above evaluations. Thus $(3) \Rightarrow (1)$.

The implication $(1) \Rightarrow (2)$ is obvious.

Problem 16.4. Show that under the conditions of Theorem 16.3 and with

$$\sup_{k>0} r(k) < 1,$$

the assertion (2) of the theorem follows from the equality

$$\lim_{u\to\infty} m_u([0,1)) = \lambda \, l([0,1)) = \lim_{u\to\infty} d_u([0,1)). \tag{16.5}$$

As the converse is also valid, (16.5) is the necessary and sufficient condition for the Poisson limit theorem.

The expressions for R and R_ε can be simplified with use of the above representation for $\varphi(u,u;r)$. It is easy to show that (4) of Theorem 16.3, follows from Berman's condition

$$r(n) \ln n \to 0, \quad n \to \infty, \tag{16.6}$$

that is, the Poisson limit theorem is valid for large values of $X(k)$ (compare this to the corresponding part of Lecture 3). Notice that from the first equality of (4) it follows that

$$u = u_N = \sqrt{2 \ln N} - \frac{\frac{1}{2} \ln \ln N + \ln(\lambda\sqrt{\pi/2})}{\sqrt{2 \ln N}} + O(1/\ln N), \quad N \to \infty, \tag{16.7}$$

where $N = [1/\Psi(u)]$. Using this, one can prove the following statement, whose complete proof can be found in Piterbarg [1996], and where the condition (16.6) is shown to be the necessary one.

Proposition 16.5. *Let* $\lim_{u\to\infty} m_u([0,1]) = \lambda$. *Then (4) of Theorem 16.3 holds if and only if for any* $\varepsilon > 0$,

$$\lim_{N\to\infty} N^{-2} \sum_{\substack{k=1, \; r(k)\ln k > \varepsilon}}^N \frac{N-k}{\sqrt{1-r(k)}} \left(\left(\frac{N}{\ln N}\right)^{\frac{2r(k)}{1+r(k)}} - 1 \right) = 0. \tag{16.8}$$

The Fisher-Tippet-Gnedenko theorem for the maximum of a Gaussian sequence follows from Theorem 16.3.

Corollary 16.6. *Let the conditions (16.8) or (16.6) hold for a stationary Gaussian sequence with zero mean and unit variance. Then*

$$P(\max_{k\in[0,n]} (X(k) - u_n)/b_n \leqslant x) \to \exp(-e^{-x})$$

as $n \to \infty$, where $b_n = \sqrt{2\ln n}$, and u_n is given by (16.7) with $\lambda = 1$.

The proof follows with the substitution $u = u_n$, $1/\Psi(u) = n$ and $\lambda = e^{-x}$ in

$$\lim_{u\to\infty} P(\xi_u([0,1] = 0) = \lim_{u\to\infty} P(\max_{k\in[0,1/\Psi(u)]} X(k) \leqslant u) = e^{-\lambda},$$

after some trivial manipulations.

17

Poisson Limit Theorem for Large Excursions of Stationary Processes

In this lecture we consider a Gaussian stationary a.s. continuous random process $X(t)$ with zero mean, unit variance, and the covariance function satisfying both Pickands' condition,

$$r(t) = 1 - |t|^\alpha + o(|t|^\alpha) \text{ as } t \to 0 \text{ for some } 0 < \alpha \leqslant 2, \tag{17.1}$$

and Berman's condition,

$$r(t) \ln t \to 0 \text{ as } t \to \infty. \tag{17.2}$$

From (17.2) it follows that for any $\delta > 0$,

$$\sup_{|t| > \delta} |r(t)| < 1. \tag{17.3}$$

Again, as in Lecture 16, we deal with the number of crossings. When $\alpha < 2$, the trajectories of X are not differentiable, so just like with the Brownian motion it is possible that the number of crossings could only be zero or infinity a.s. Nonetheless, in order to be able to talk about the number of crossings, let us introduce a concept of the a-point of crossing. We say that a point t is an a-point of up-crossing of a level u by a trajectory $X(t,\omega)$ if $X(t,\omega) = u$ and $X(s,\omega) < u$ for all $s \in [t-a,t)$. Our aim in this lecture is to prove the Poisson limit theorem for the random point process of a-points of up-crossings of a high (i.e. approaching infinity) level. Let us denote

$$P_X(u, W) = P\left(\max_{t \in W} X(t) \leqslant u\right) \text{ and } \bar{P}_X(u, W) = 1 - P_X(u, W),$$

Consider the function

$$\mu(u) = H_\alpha u^{2/\alpha} \Psi(u).$$

Recall that $\Psi(u) = (\sqrt{2\pi}u)^{-1} e^{-u^2/2}$, H_α is the Pickands constant, $0 < H_\alpha < \infty$ (see Lecture 9 for the definition); and by Theorem 9.15 with (17.3), for any T,

$$\bar{P}_X(u,[0,T]) = T\mu(u)(1 + o(1)) \text{ as } u \to \infty. \qquad (17.4)$$

Let $\eta_{a,u}(B)$, $B \in \mathcal{B}$, where $\mathcal{B}$ is the σ-ring of bounded Borel sets on $\mathbb{R}$, be the point process of a-points of up-crossings of u by $X(t)$. Its intensity tends to zero as $u \to \infty$, therefore, following the arguments of Lecture 16, let us introduce the normalized point process

$$\Phi_u(B) = \eta_{a,u}(\mu(u)^{-1}B)$$

and denote by $\pi(\cdot)$ the Poisson stationary point process on $\mathcal{B}$ with intensity one.

Theorem 17.1. *Consider a Gaussian stationary process $X(t)$ with zero mean and unit variance such that (17.1) and (17.2) are satisfied. Then for any a, $a > 0$, the random point processes $\Phi_u(B)$, $B \in \mathcal{B}$, converge weakly as $u \to \infty$ to the Poisson stationary point process $\pi(B)$, $B \in \mathcal{B}$.*

By Theorem 16.2, to prove the theorem it is sufficient to verify Kallenberg's conditions,

$$\forall L \in \mathcal{L}, \ \lim_{u \to \infty} P(\Phi_u(L) = 0) = P(\pi(L) = 0) \text{ and } \limsup_{u \to \infty} E\Phi_u(L) \leqslant E\pi(L)$$
$$(17.5)$$

for the sub-ring $\mathcal{L}$ of $\mathcal{B}$ generated by all finite intervals $[a,b)$. Notice that the property $\pi(\partial L) = 0$ a.s. is obvious.

Similarly to Lecture 16, by the standard arguments, considering $X(t)$ on an infinitely increasing interval $[0,T]$, $T \to \infty$, and choosing $u = u(T)$ so that $T\mu(u_T) \to e^{-x}$ as $T \to \infty$, one can obtain from Theorem 17.1 the limit theorem for maximal values of $X(t)$:

Corollary 17.2. *Let (17.1) and (17.2) be fulfilled, then*

$$\lim_{T \to \infty} P\left(\max_{t \in [0,T]} X(t) \leqslant \frac{x}{a_T} + b_T\right) = e^{-e^{-x}}$$

with

$$a_T = \sqrt{2 \ln T}, \ b_T = \sqrt{2 \ln T} + \frac{\left(\frac{1}{\alpha} - \frac{1}{2}\right) \ln(2 \ln T) + \ln\left(\frac{H_\alpha}{\sqrt{2\pi}}\right)}{\sqrt{2 \ln T}}.$$

Proof of Theorem 17.1.

Lemma 17.3. *For $\mu_a(u)$, the intensity of $\eta_{a,u}(B)$,*

$$\lim_{u \to \infty} \mu_a(u)/\mu(u) = 1.$$

Proof. The interval $I = [0, a/2)$ cannot contain more than one a-point of up-crossing. Therefore it is sufficient to evaluate asymptotic behavior of

$$\mathrm{P}\left(\eta_{a,u}(I) > 0\right) = \mathrm{E}\eta_{a,u}(I) = \frac{a}{2}\mu_a(u).$$

as $u \to \infty$. Notice that $\{\eta_{a,u}(I) > 0\} \subset \{\max_{t\in I} X(t) > u\}$. On the other hand,

$$\{\eta_{a,u}(I) > 0\} \supset \left\{\max_{t\in(-a,0)} X(t) \leqslant u, \max_{t\in I} X(t) > u\right\}. \tag{17.6}$$

Moreover,

$$\mathrm{P}\left\{\max_{t\in(-a,0)} X(t) \leqslant u, \max_{t\in I} X(t) > u\right\}$$

$$= \bar{P}_X(u, I) - \mathrm{P}\left(\max_{t\in I} X(t) > u, \max_{(-a,0)} X(t) > u\right)$$

$$= \bar{P}_X(u, I) - \left[\bar{P}_X(u, I) + \bar{P}_X(u, (-a, 0)) - \bar{P}_X(u, (-a, a/2))\right].$$

We can now make use of Picands' theorem. The expression in the square brackets is infinitely smaller than $\bar{P}_X(u, I)$, which proves the lemma.

Lemma 17.4. *For any $L \in \mathcal{L}$,*

$$\mathrm{P}\left(\Phi_u(L) = 0\right) = P_X(u, \mu(u)^{-1}L) + o(1)$$

as $u \to \infty$.

Proof. Denote $L_u = \mu(u)^{-1}L$, $L \in \mathcal{L}$. We have, $P_X(u, L_u) \leqslant \mathrm{P}\left(\Phi_u(L) = 0\right)$. Furthermore, similarly to (17.6),

$$\mathrm{P}\left(\Phi_u(L) = 0\right) \leqslant P_X(u, L_u) + \mathrm{P}\left(\max_{t\in L_u} X(t) > u, \max_{(L_u\oplus(-a,0))\setminus L_u} X(t) > u\right),$$

where $\oplus$ is the Minkowski sum of sets, $A \oplus B = \{t + s : t \in A, s \in B\}$. Since L consists of a finite number of disjoint intervals, we have,

$$\mathrm{P}\left(\max_{t\in L_u} X(t) > u, \max_{(L_u\oplus(-a,0))\setminus L_u} X(t) > u\right) \leqslant \mathrm{P}\left(\max_{(L_u\oplus(-a,0))\setminus L_u} X(t) > u\right)$$

$$= O\left(u^{2/\alpha}\Psi(u)\right) = o(1)$$

as $u \to \infty$.

In order to prove the Poisson limit theorem we, again, are going to use the comparison result of Lecture 2. To this end, let us consider the grids on $\mathbb{R}$,

$$\mathcal{R}_b := \{bku^{-2/\alpha}, k \in \mathbb{Z}\}, \ b > 0.$$

Lemma 17.5. *For any $L \in \mathcal{L}$ and an arbitrary $\epsilon > 0$, one can find $b > 0$, $u_0 > 0$, such that for all $u \geqslant u_0$,*

$$P_X(u, L_u \cap \mathcal{R}_b) - P_X(u, L_u) \leqslant \mathrm{Diam}(L)\epsilon,$$

Proof. We have,

$$P_X(u, L_u \cap \mathcal{R}_b) - P_X(u, L_u) = P\left(\max_{L_u \cap \mathcal{R}_b} X(t) \leqslant u, \max_{L_u} X(t) > u\right).$$

By stationarity, we can consider $L \in \mathcal{L}$ only on $\mathbb{R}_+$. Let $T = \mathrm{Diam}(L)$ and $L \subset [0, T]$. Let us cover $[0, \mu(u)^{-1}T)$ with disjoint intervals of length $\lambda u^{-2 \cdot \alpha}$. The number of such small intervals is equal to $[\mu(u)^{-1}u^{2/\alpha}T] + 1$. By stationarity,

$$P\left(\max_{L_u \cap \mathcal{R}_b} X(t) \leqslant u, \max_{L_u} X(t) > u\right) \tag{17.7}$$

$$\leqslant \left([\mu(u)^{-1}u^{2/\alpha}T] + 1\right) P\left(\max_{[0, u^{-2/\alpha}] \cap \mathcal{R}_b} X(t) \leqslant u, \max_{[0, u^{-2/\alpha}]} X(t) > u\right).$$

Repeating word-for-word the proof of Lemma 9.1, we obtain that

$$\lim_{u \to \infty} \Psi(u)^{-1} P\left(\max_{[0, u^{-2/\alpha}] \cap \mathcal{R}_b} X(t) \leqslant u, \max_{[0, u^{-2/\alpha}]} X(t) > u\right)$$

$$= \int_0^\infty e^s P\left(\max_{t \in [0,1] \cap b\mathbb{Z}} \sqrt{2}B_{\alpha/2}(t) - t^\alpha \leqslant s, \max_{t \in [0,1]} \sqrt{2}B_{\alpha/2}(t) - t^\alpha > s\right) ds.$$

Since the trajectories of the fractional Brownian motion are a.s. continuous, the probability under the integral tends to zero for all s, as $b \to 0$. By the dominating convergence, the integral tends to zero as well. Let us chose b so small that the integral does not exceed $\varepsilon/2$. Now, with such b and a sufficiently large u_0, the lemma follows from (17.7).

Let us continue the proof of the theorem, considering again the interval $[0, T]$ covering L. Let us take $\delta > 0$ and split the interval $[0, \mu(u)^{-1}T]$ into intervals of unit length, calling them long intervals, with intervals of length δ called short intervals. In both cases the last interval can be shorter than 1 or δ. Let us denote by N the number of the long intervals, and by Π_u their union. Likewise, let us denote by Λ_u the union of all short intervals.

Lemma 17.6. *For any $L \in \mathcal{L}$ with positive length and any $\varepsilon > 0$ there exists $\delta > 0$ such that for all sufficiently large u,*

$$P_X(u, \Pi_u \cap \mathcal{R}_b) - P_X(u, L_u \cap \mathcal{R}_b) \leqslant \varepsilon.$$

Proof. By (17.4), for all sufficiently large u,

$$0 \leqslant P_X(u, \Pi_u \cap \mathcal{R}_b) - P_X(u, L_u \cap \mathcal{R}_b)$$

$$= P\left(\max_{\Pi_u \cap \mathcal{R}_b} X(t) \leqslant u, \max_{L_u \cap \mathcal{R}_b} X(t) > u\right)$$

$$\leqslant \bar{P}_X(u, L_u \backslash \Pi_u) \leqslant 2\frac{\mu(u)^{-1}T}{1+\delta}\bar{P}_X(u, [0, \delta])$$

$$\leqslant \frac{3\mu(u)^{-1}T}{1+\delta}\delta\mu(u) = \frac{3\delta T}{1+\delta} =: \varepsilon,$$

with $\delta = \varepsilon/(3T - \varepsilon)$ and $T = \mathrm{Diam}(L)$.

Let us denote by K_j, $j = 1, 2, \ldots$, the long intervals that are the summands in Π_u. Recall that these long intervals alternate with the short ones, of length δ. Let us introduce the collection of independent copies of $X(t)$, $X_j(t)$, $t \in K_j$, $j = 1, 2, \ldots$, and the Gaussian random process,

$$X_0(t) = X_j(t) \text{ if } t \in K_j, \ j = 1, 2, \ldots.$$

Lemma 17.7. *For all $L \in \mathcal{L}$,*

$$P_X(u, \Pi_u \cap \mathcal{R}_b) - P_{X_0}(u, \Pi_u \cap \mathcal{R}_b) \to 0$$

as $u \to \infty$.

Proof. With $S(i, j) := \{s, t : s \in K_j \cap \mathcal{R}_b, t \in K_i \cap \mathcal{R}_b\}$, by the comparison inequality of Corollary 2.8,

$$|P_X(u, \Pi_u \cap \mathcal{R}_b) - P_{X_0}(u, \Pi_u \cap \mathcal{R}_b)|$$

$$\leqslant \frac{1}{\pi} \sum_{s,t \in \lambda_u \cap \mathcal{R}_b, s \neq t} |r(t - s) - r_0(s, t)|$$

$$\times \int_0^1 (1 - r_h(s, t))^{-1/2} \exp\left(-\frac{u^2}{1 + r_h(s, t)}\right) dh$$

$$= \frac{1}{\pi} \sum_{i \neq j} \sum_{S(i,j)} |r(t - s)| \int_0^1 (1 - hr(t - s))^{-1/2} \exp\left(-\frac{u^2}{1 + hr(s, t)}\right) dh,$$

where

$$r_h(s, t) = hr(s - t) + (1 - h)r_0(s, t),$$

$s, t \in \mathbb{R}$, and r_0 is the covariance function of X_0. Let us prove that the last sum tends to zero. We have,

$$\gamma_2 := \sup_{t \geqslant \delta}(1 - |r(t)|) \in (0, 1],$$

Let us consider first those long intervals that are not too far away from each other. Namely, defining the distance

$$d(K_i, K_j) := \sup\{|t - s| : t \in K_i, s \in K_j\}$$

and taking $\gamma_1 \in (0, \gamma_2/2)$ we consider the part Σ_1 of the sum on $s \in K_j$, $t \in K_i$ with $d(K_i, K_j) \leqslant \mu(u)^{-\gamma_1}$. With such γ_2 and some constants C_1 C_2, we have,

$$\Sigma_1 \leqslant C_1 \sum_{\substack{i \neq j \\ d(K_i, K_j) \leqslant \mu(u)^{-\gamma_1}}} \sum_{S(i,j)} \exp\left(-\frac{u^2}{1 + |r(t-s)|}\right)$$

$$\leqslant C_2 \sum_{\substack{i \neq j \\ d(K_i, K_j) \leqslant \mu(u)^{-\gamma_1}}} \sum_{S(i,j)} \exp\left(-\frac{u^2}{2 - \gamma_2}\right)$$

$$= O\left(\mu(u)^{-1}\mu(u)^{-\gamma_1} u^{n-1+2/\alpha} e^{-u^2/(2-\gamma_2)}\right) = o(1).$$

as $u \to \infty$.

Now let us turn to s, t with $d(K_i, K_j) \geqslant \mu(u)^{-\gamma_1}$, $t \in K_i$, $s \in K_j$, denoting the corresponding sum by Σ_2. Making use (17.2),

$$\sup_{|t-s| \geqslant \mu(u)^{-\gamma_1}} r(t - s) =: \kappa(u) = o(u^{-2})$$

as $u \to \infty$, that is, for some C_3, C_4

$$\Sigma_2 \leqslant C_3 \kappa(u) \sum_{\substack{i \neq j \\ d(K_i, K_j) \geqslant \mu(u)^{-\gamma_1}}} \sum_{S(i,j)} \exp\left(-\frac{u^2}{1 + |r(t-s)|}\right)$$

$$\leqslant C_4 \kappa(u) e^{-u^2} \sum_{\substack{i \neq j \\ d(K_i, K_j) \geqslant \mu(u)^{-\gamma_1}}} \sum_{S(i,j)} \exp\left(\frac{|r(t-s)|u^2}{1 + |r(t-s)|}\right)$$

$$= O\left(\left(\mu(u)^{-1}u^{2/\alpha}\right)^2 \kappa(u)e^{-u^2}\right) = O(u^2\kappa(u)) = o(1)$$

as $u \to \infty$.

Notice now that the assertion of Lemma 17.6 is valid for X_0 as well, with the same grid. For any $\varepsilon > 0$ there exists δ such that $|N\mu(u) - |L|| \leqslant \varepsilon$, $|L|$ is the length of L, recall that N is the number of K_j from Π_u. Using this, as X_0 consists if independent parts on different K_j, we have,

$$P_{X_0}(u, \Pi_u \cap R_b)) = (1 - \overline{P}_{X_0}(u, [0,1) \cap R_b))^N \to e^{-|L|}.$$

The first statement in (17.5) now follows from this result and Lemmas 17.3 – 17.7. The second statement in (17.5) follows by the equivalence of $\mu(u)$ and $\mu_a(u)$, (Lemma 17.3), that is, $\lim_{u \to \infty} E\Phi_u(L) = E\pi(L)$. This completes the proof of Theorem 17.1.

Generalizations of this theorem to Gaussian fields and other related results can be found in Piterbarg [1996].

Comparison Identity and Inequality in Continuous Time

In this lecture we consider generalizations of the comparison identity and Berman's comparison inequality to stationary Gaussian processes in continuous time. The following corollary to Theorem 2.6 will be useful.

Theorem 18.1. *Let* $\mathbf{X}_0 = (X_{01},\ldots,X_{0d})$ *and* $\mathbf{X}_1 = (X_{11},\ldots,X_{1d})$ *be two independent Gaussian vectors with zero means and the covariances* r_{0ij} *and* r_{1ij}, *respectively, with* $r_{0ii} = r_{1ii} = 1$ *for all* $i = 1,\ldots,d$. *Assume that* $|r_{0ij}| < 1$ *for all* $i \neq j$. *Then for any* u,

$$P(\max_{i=1,\ldots,d} X_{1i} \leqslant u) - P(\max_{i=1,\ldots,d} X_{0i} \leqslant u) = \sum_{k,l=1,\ k>l}^{d} (r_{1kl} - r_{0kl})$$

$$\times \int_0^1 \varphi(u,u;r_{hkl})P(\max_{i=1,\ldots,d} X_{hi} \leqslant u \mid X_{hk} = X_{hl} = u)dh, \quad (18.1)$$

where $\mathbf{X}_h = \sqrt{h}\mathbf{X}_1 + \sqrt{1-h}\mathbf{X}_0$ *and* $\varphi(u,u;r)$ *is the two-dimensional Gaussian density with zero means, unit variances and the covariance* r.

The idea of this generalization is simple and similar to how the formulas for the moments of the number of crossings were derived. We first write down the above comparison identity on a grid and then try to pass to the limit on the right-hand side of (18.1) in the right way, as the step of the grid goes to zero. The left-hand side converges to the estimated difference of maximum distributions by continuity. This "right-way" passage to the limit on the right-hand side is the main part of the proof of the following comparison theorem.

Let us denote by $\varphi_{\mathbf{Z}}(\mathbf{z})$ the probability density of a Gaussian vector $\mathbf{Z}$.

Theorem 18.2. *Consider two independent Gaussian twice-differentiable in square mean stationary processes* $X_\nu(t)$, $\nu = 0,1$, $t \in [0,T)$, $T \leqslant \infty$, *with a.s. continuous trajectories and zero means. Assume that they satisfy the* *conditions* $EX_v(t)^2 \equiv EX'_v(t)^2 \equiv 1, v = 0,1.$ *Let the distributions of Gaussian vectors*

$$(X_v(0), X_v(t), X'_v(0), X'_v(t), X''_v(0), X''_v(t)), \quad v = 0,1,$$

be non-degenerate for all $t > 0$. Denote

$$X_h(t) = \sqrt{h}X_1(t) + \sqrt{1-h}X_0(t), \quad h \in [0,1].$$

Then, for any $S < T$ and all u,

$$P(\max_{t\in[0,S]} X_1(t) \leq u) - P(\max_{t\in[0,S]} X_0(t) \leq u)$$

$$= \int_0^1 dh \left\{ \int_0^S t\,dt(r_1(t) - r_0(t))\varphi_{X_h(0),X_h(t),X'_h(0),X'_h(t)}(u,u,0,0) \right.$$

$$\times \int_{-\infty}^0 dy_1 \int_{-\infty}^0 dy_2 y_1 y_2 \varphi_{X''_h(0),X''_h(t)}(y_1,y_2 \mid u,u,0,0)$$

$$\times P(\max_{s\in[0,S]} X_h(s) \leq u \mid u,u,0,0,y_1,y_2)$$

$$+2\int_0^S dt(r_1(t) - r_0(t))\varphi_{X_h(0),X_h(t),X'_h(t)}(u,u,0)$$

$$\times \int_{-\infty}^0 dy|y|\varphi_{X''_h(t)}(y \mid u,u,0)P(\max_{s\in[0,S]} X_h(s) \leq u \mid u,u,0,y)$$

$$\left. +(r_1(S) - r_0(S))\varphi_{X_h(0),X_h(S)}(u,u)P(\max_{s\in[0,S]} X_h(s) \leq u \mid u,u) \right\}, \quad (18.2)$$

where r_v, $v = 0,1$, are the covariance functions of X_v, $v = 0,1$, respectively; $\varphi_{X''_h(0),X''_h(t)}(y_1,y_2 \mid u,u,0,0)$ is the conditional probability density of $(X''_h(0), X''_h(t))$ given

$$(X_h(0), X_h(t), X'_h(0), X'_h(t)) = (u,u,0,0);$$

$\varphi_{X''_h(t)}(y \mid u,u,0)$ *is the conditional probability density of $X''_h(t)$ given*

$$(X_h(0), X_h(t), X'_h(t)) = (u,u,0).$$

The conditions in all the conditional probabilities are the values of the corresponding random variables from the same integral. In the last integral, the condition is the value of $(X_h(0), X_h(S))$.

Proof. Let us rewrite the identity (18.1) on a time grid,

$$P(\max_{0\leq k\leq 2^n S} X_1(k2^{-n}) \leq u) - P(\max_{0\leq k\leq 2^n S} X_0(i2^{-n}) \leq u)$$

$$= \sum_{i<j}(r_1((j-i)2^{-n}) - r_0((j-i)2^{-n})) \int_0^1 \varphi(u,u)$$

$$\times P(\max_{0\leq k\leq 2^n S} X_h(k2^{-n}) \leq u \mid X_h(i2^{-n}) = X_h(j2^{-n}) = u)dh, \quad (18.3)$$

where $\varphi(u,u)$ is the probability density of $(X_h(l2^{-n}), X_h(k2^{-n}))$, $r_h = hr_1 + (1-h)r_0$. Assume for a start that all finite-dimensional probability densities exist for both processes. Let us proceed with the conditional distributions under the integral, denoting them by P_{ij}. Let us further denote

$$t_1 := 2^{-n}i, \ t_2 := 2^{-n}j, \ A := \{ \max_{0 \leqslant i \leqslant 2^n S} X_h(i2^{-n}) \leqslant u \},$$

and consider initially the case $t_1 > 0$, $t_2 < S$, that is, both points are the inner points of $[0,S]$. By Tailor's formula, we have for $l = 1,2$,

$$X_h(t_l \pm 2^{-n}) = X_h(t_l) \pm 2^{-n}X'_h(t_l) + \frac{1}{2}2^{-2n}X''_h(t_l) + 2^{-2n}\alpha^{\pm}_{nl}\xi^{\pm}_{nl}, \qquad (18.4)$$

where the four deterministic sequences $\alpha^{\pm}_{nl}$ tend to zero as $n \to \infty$, and $\xi^{\pm}_{nl}$, $l = 1,2$, $n = 1,2,\ldots$ are four jointly Gaussian sequences with zero means and unit variances. Now consider the event

$$A_{t_1 t_2} = A \cap \{2^{-n-1}X''_h(t_l) + 2^{-n}\alpha^+_{nl}\xi^+_{nl} \leqslant X'_h(t_l)$$
$$\leqslant -2^{-n-1}X''_h(t_l) + 2^{-n}\alpha^-_{nl}\xi^-_{nl}, \ l = 1,2\}.$$

We have, using $X_h(t_l) = u$, $l = 1,2$,

$$\begin{aligned}
P_{ij} &= P(A_{t_1 t_2} \mid X_h(t_l) = u, l = 1,2) \\
&= E[E[P(A_{t_1 t_2} \mid X_h(t_l) = u, l = 1,2)|X''_h(t_l), \xi^{\pm}_{nl}, l = 1,2]]. \qquad (18.5)
\end{aligned}$$

Let us denote

$$N_{ij} = \{k \in \{0,\ldots,2^n S\}\backslash\{i \pm 1, j \pm 1\}\},$$
$$S_{ij} = \{\mathbf{z} = (z_k, k \in N_{ij}) : z_k \leqslant u, k \in N_{ij}\}.$$

Then

$$P_{ij} = \int_{S_{ij}} d\mathbf{z} \int_{2^{-n-1}X''_h(t_1)+2^{-n}\alpha^+_{n1}\xi^+_{n1}}^{-2^{-n-1}X''_h(t_1)+2^{-n}\alpha^-_{n1}\xi^-_{n1}} dx_1$$
$$\int_{2^{-n-1}X''_h(t_2)+2^{-n}\alpha^+_{n2}\xi^+_{n2}}^{-2^{-n-1}X''_h(t_2)+2^{-n}\alpha^-_{n2}\xi^-_{n2}} dx_2 \varphi_h(\mathbf{z},x_1,x_2), \qquad (18.6)$$

where φ_h is the conditional probability density of

$$(X_h(k2^{-n}), k \in N_{ij}, X'_h(t_l), l = 1,2)$$

given $X_h(t_l) = u$, $l = 1,2$. Substituting $x_l = 2^{-n}x'_l$, $l = 1,2$, we obtain,

$$P_{ij} = 2^{-2n} \int_{S_{ij}} d\mathbf{z} \int_{2^{-1}X''_h(t_1)+\alpha^+_{n1}\xi^+_{n1}}^{-2^{-1}X''_h(t_1)+\alpha^-_{n1}\xi^-_{n1}} dx'_1$$
$$\int_{2^{-1}X''_h(t_2)+\alpha^+_{n2}\xi^+_{n2}}^{-2^{-1}X''_h(t_2)+\alpha^-_{n2}\xi^-_{n2}} dx'_2 \varphi_{1h}(\mathbf{z} \mid 2^{-n}x'_l, l = 1,2)\varphi_{2h}(2^{-n}x'_l, l = 1,2), \qquad (18.7)$$

where φ_{1h} is the conditional probability density of $(X_h(k2^{-n}), k \in N_{ij})$ given

$$X_h(t_l) = u, X'_h(t_l) = 2^{-n}x'_l, \ l = 1, 2;$$

and φ_{2h} is the conditional probability density of $(X'_h(t_l), l = 1, 2)$ given $X_h(t_l) = u, l = 1, 2$. The relation (18.7) can be written as

$$P_{ij} = 2^{-2n} \left(\int_{S_{ij}} d\mathbf{z} \int_{2^{-1}X''_h(t_1)}^{-2^{-1}X''_h(t_1)} dx'_1 \right.$$

$$\left. \int_{2^{-1}X''_h(t_2)}^{-2^{-1}X''_h(t_2)} dx'_2 \, \varphi_{1h}(\mathbf{z} \,|\, 2^{-n}x'_l, l = 1, 2) \varphi_{2h}(2^{-n}x'_l, l = 1, 2) + R_n \right). \quad (18.8)$$

The last probability density has its maximum at zero, therefore, subtracting (18.8) from (18.7), we obtain that

$$|R_n| \leqslant \varphi_{2h}(0,0)(|\alpha_{n1}^- \xi_{n1}^-| + |\alpha_{n1}^+ \xi_{n1}^+| + |\alpha_{n2}^- \xi_{n2}^-| + |\alpha_{n2}^+ \xi_{n2}^+|). \quad (18.9)$$

Since all the probability densities exist and the conditional probability densities are continuous with respect to the conditions, the dominating convergence theorem can be applied to the right-hand side of (18.8), with fixed t_1, t_2. Therefore

$$P_{ij} = 2^{-2n} \left(P(\max_{[0,S]} X_h(s) \leqslant u \mid X_h(t_l) = u, X'_h(t_l) = 0, l = 1, 2) \right.$$

$$\left. \times (X''_h(t_1)_-)(X''_h(t_1)_-) \varphi_{2h}(0,0) + o(1) \right), \quad (18.10)$$

a.s., where $a_- = \min(a, 0)$. By virtue of (18.9), from (18.5) it follows for fixed t_1, t_2 that

$$P_{ij} = 2^{-2n} \left(\int_{-\infty}^{0} \int_{-\infty}^{0} y_1 y_2 P_{t_1 t_2}(u, y_1, y_2) \varphi_{2h}(0, 0, y_1, y_2) dy_1 dy_2 + o(1) \right), \quad (18.11)$$

as $n \to \infty$. Here

$$P_{t_1 t_2}(u, y_1, y_2)$$
$$= P(\max_{[0,S]} X_h(s) \leqslant u \mid X_h(t_l) = u, X'_h(t_l) = 0, X''_h(t_l) = y_l, l = 1, 2),$$

and φ_{2h} is the conditional probability density of

$$(X'_h(t_l), l = 1, 2, X''_h(t_l), l = 1, 2)$$

given $X_h(t_l) = u, l = 1, 2$.

The other sums in (18.3) with $i = 0$ ($t_1 = 0$), $j < 2^n S$ ($t_2 < S$) and with $i > 0$ ($t_1 > 0$), $j = 2^n S$ ($t_2 = S$) can be evaluated similarly. The Tailor expansion (18.4) is applied at only one point. Conditional probability densities will differ from the ones considered above as they correspond to the other members in the comparison identity (18.2). For example, in the first case, $t_1 = 0, t_2 < S$, we would have,

$$P_{0j} = 2^{-n} \left(\int_{-\infty}^{0} \int_{-\infty}^{0} |y| P_{t_2}(u, y) \varphi_{3h}(0, y) dy + o(1) \right), \qquad (18.12)$$

as $n \to \infty$, where

$$P_{t_2}(u, y) = P(\max_{[0,S]} X_h(s) \leqslant u \mid X_h(t_l) = u, l = 1, 2, X_h'(t_2) = 0, X_h''(t_2) = y),$$

and φ_{3h} is the conditional probability density of $(X_h'(t_2), X_h''(t_2))$ given $X_h(t_l) = u$, $l = 1, 2$. At last, the evaluation of the summand with $i = 0$, $(t_1 = 0)$, and $j = 2^n S$ $(t_2 = S)$, as $n \to \infty$ is trivial.

All that remains is to sum the limits of terms on the right-hand side of (18.3). It is clear that one should split the sum onto four parts: the sum over the inner points, two sums with one of the end points fixed, and the term with both end points. Since by our temporal assumption all the densities exist, there are no problems with summing over $|t_2 - t_1| \geqslant \varepsilon$, with any positive ε, to obtain the integrals on the right-hand side of (18.2). In order to pass from ε to 0, let us take $X_0^{\delta}(t) = X_0((1 - \delta)t)$ instead of $X_0(t)$, where $\delta > 0$ is sufficiently small. Then, since the first and second derivatives of the covariance functions of X_0 and X_1 coincide, we have for sufficiently small t that $r_0((1-\delta)t) \geqslant r_1(t)$. Now let us drop from the right-hand side of (18.1) the terms with t_1, t_2 that satisfy this inequality. Then (18.2) becomes an estimate from above, where we make use of the monotone convergence theorem to let ε go to zero and get an estimate from above for any sufficiently small δ. Then we let $\delta \downarrow 0$, again using the monotone convergence theorem and to obtain the "$\leqslant$" inequality in (18.2). In order to prove the inequality "$\geqslant$", let us now consider $X_0^{\delta}(t) = X_0((1+\delta)t)$ and repeat the argument above.

Now it remains to drop our assumption on the existence of densities. We already know how to do this: just add to each of the processes independent copies Z_1, Z_2 of a smooth-enough Gaussian stationary process $Z(t)$ with non-degenerate finite-dimensional distributions. Namely, we could set

$$X_{\kappa,i}(t) = \sqrt{\kappa} Z_i(t) + \sqrt{1 - \kappa} X_i(t), \quad i = 0, 1,$$

write down the identity (18.2) for them, and then let $\kappa \to 0$.

To use Theorem 18.2 in practice we should derive an upper bound for the difference of probabilities, estimating from above the right-hand side of (18.2) by a quantity of the order $\exp(-u^2/1 + \rho))$, $\rho < 1$. We already did this in discrete time. In the present situation, with continuous time, deriving such an estimate is technically much more challenging. Let us outline here the main steps, in order to understand the order of the upper bound. The reader can fill in the details of the derivation himself or read about it in Piterbarg [1982]. For Gaussian fields one can look in Piterbarg [1996].

Let us denote

$$m_1(t) = E(X_h''(t)|X_h(t) = X_h(0) = 1, X_h'(t) = X_h'(0) = 0),$$
$$\sigma_1^2(t) = \text{Var}(X_h''(t)|X_h(t) = X_h(0) = 1, X_h'(t) = X_h'(0) = 0),$$

$$m_2(t) = E(X_h''(t)|X_h(t) = X_h(0) = 1, X_h'(t) = 0),$$
$$\sigma_2^2(t) = \text{Var}(X_h''(t)|X_h(t) = X_h(0) = 1, X_h'(t) = 0).$$

Estimating the probabilities on the right-hand side of (18.2) by one, and the integrals over y_1, y_2, by the conditional moments of $X_h''(t)$, we get the following result.

Theorem 18.3. *Let the conditions of Theorem 18.2 be fulfilled. Then*

$$|P(\max_{t \in [0,S]} X_1(t) \leqslant u) - P(\max_{t \in [0,S]} X_0(t) \leqslant u)|$$

$$\leqslant \int_0^1 \int_0^S |r_1(t) - r_0(t)|$$

$$\times [t(\sigma_1(t) + u|m_1(t)|)^2 \varphi_{X_h(0),X_h(t),X_h'(0),X_h'(t)}(u,u,0,0)$$

$$+ 2(\sigma_2(t) + u|m_2(t)|)\varphi_{X_h(0),X_h(t),X_h'(t)}(u,u,0)]dhdt$$

$$+ |r_1(S) - r_0(S)| \int_0^1 \varphi_{X_h(0),X_h(S)}(u,u)dh. \qquad (18.13)$$

It is the easiest to estimate $\sigma_1(t)$ and $\sigma_2(t)$, since the conditional variances cannot be higher than the respective unconditional ones,

$$\max_{i=1,2} \sigma_i^2(t) \leqslant \max_{i=0,1}(EX_i''(0))^2.$$

Writing down the expressions for conditional expectations and making use of some calculus for studying the behavior of $m_i(t)$ at zero, one can get that for some constant M, $\sup_{t \geqslant 0, i=1,2} m_i(t) \leqslant M$.

In the next steps one has to be a bit more precise to get the required exponential behavior.

Lemma 18.4. *Under the assumptions of Theorem 18.2, for the conditional density of $(X_h(0), X_h(t))$ given $X_h'(0) = X_h'(t) = 0$,*

$$\varphi_{X_h(0),X_h(t)}(u,u \mid X_h'(0) = X_h'(t) = 0)$$

$$= \frac{\exp(-u^2/(1 + R_h(t)))}{2\pi\sqrt{(1 + R_h(t))(1 - Q_h(t))}}, \qquad (18.14)$$

where

$$R_h(t) = r_h(t) - \frac{r_h'(t)^2}{1 + r_h''(t)}, \quad Q_h(t) = r_h(t) + \frac{r_h'(t)^2}{1 - r_h''(t)}.$$

Now let us expand $R_h(t)$ at zero with Tailor's formula to get

$$\lim_{t\to 0} R_h(t) = 1 - 2EX_h''(0)^2,$$

that is,

$$1 + R_h(0) = 2(1 - EX_h''(0)^2) < 2.$$

From here it also follows that

$$\max_{h\in[0,1],t\in[0,\varepsilon]} 1 + R_h(t) < 2$$

for some $\varepsilon > 0$. Furthermore, for $t > \varepsilon$, we have that $1 + R_h(t) \leqslant 1 + r_h(t) < 1 + \sup_{t\geqslant\varepsilon} \max(r_1(t), r_0(t)) < 2$. It is recommended that the reader estimate the denominator in (18.14), as well as the probability density of $(X_h'(0), X_h'(t))$ at $(0,0)$, by himself.

Lemma 18.5. *Under the assumption of Theorem 18.2, for the probability density of $(X_h(0), X_h(t), X_h'(t))$,*

$$\varphi_{X_h(0), X_h(t), X_h'(t)}(u, u, 0) = \frac{\exp\left(-\frac{1}{2}u^2\left(1 + (1 - r_h(t))^2/d(h)\right)\right)}{\sqrt{(2\pi)^3 d_h(t)}},$$

where $d_h(t) = 1 - r_h^2(t) - r_h'(t)^2$. Moreover,

$$\lim_{t\to 0} t^{-4} d_h(t) = \frac{1}{4}(EX_h''(0)^2 - 1) > 0.$$

Notice that the last limit is positive, otherwise $X(t)$ would be a sinusoid, and some of our densities would not exist.

To proceed, this scheme should be repeated for other integrals as well. It should not be difficult for the reader to do that. At last, the last summand in (18.13) can also be easily estimated by the reader. Thus we get the following.

Theorem 18.6. *Let two Gaussian stationary processes $X_v(t)$, $v = 0,1$, $t \in [0,T)$, $T \leqslant \infty$, satisfy the conditions of Theorem 18.2 except for their independence. Then for any $t_0 > 0$ there exist $\rho > 0$ and C such that for all $S < T$ and u,*

$$\left| P\left(\max_{t\in[0,S]} X_1(t) \leqslant u\right) - P\left(\max_{t\in[0,S]} X_0(t) \leqslant u\right) \right|$$
$$\leqslant C\left[S\exp\left(-\frac{(1+\delta)u^2}{2}\right) \right.$$
$$\left. + S\int_{t_0}^{S} |r_0(t) - r_1(t)| \exp\left(-\frac{u^2}{1+\widehat{r}(t)}\right) dt \right], \tag{18.15}$$

where $\widehat{r}(t) = \max(r_0(t), r_1(t))$.

From the structure of the right-hand side of this inequality it is easily to guess that in order to prove it, one should first split the integrals in (18.2) into two parts: from zero to t_0, and from t_0 to S. This form of the inequality allows one to use the fact that $r_i(t)$ are small for large t.

19

On Rate of Convergence in the Limit Theorem for the Maximum

In this lecture we show that the rate of convergence for the number of high excursions of a Gaussian process in the Poisson limit theorem can be very slow; specifically, it can be logarithmic. In addition, for the case of smooth Gaussian processes, we suggest a correction to the Poisson limit theorem that gives a power rate of convergence.

Let us formulate a modification of Theorem 15.2 that we will use later.

Theorem 19.1. *Let $X(t)$ be a Gaussian stationary process with zero mean and unit variance, whose covariance function $r(t)$ is twice differentiable. Assume that*

$$EX(t)^2 = EX'(t)^2 = 1 \tag{19.1}$$

and

$$E(X'(t) - X'(0))^2 = 2(r''(t) - r''(0)) \leqslant G|t|^2. \tag{19.2}$$

Then there exist positive Δ and C such that for any T,

$$P(\max_{t \in [0,T]} X(t) > u) = \frac{T}{2\pi} e^{-u^2/2} + 1 - \Phi(u) + R(u),$$

where

$$0 \leqslant R(u) \leqslant CT^2 \exp\left(-\frac{(1+\Delta)u^2}{2}\right).$$

Another important tool that we will use is the comparison theorem, Theorem 18.6. We apply the theorem to $X_0(t)$ that we choose to be a Gaussian stationary process with a sufficiently smooth covariance function r_0 that has compact support and, moreover, satisfies $r_0(0) = -r_0''(0) = 1$.

Problem 19.2. Show that such a covariance function exists.

Hint: Consider a random process as a stochastic integral

$$X(t) = \int \psi(s+t) dB_s$$

with ψ, the test function, infinitely differentiable with compact support.

Let $t_0 = \text{Diam}(\text{supp}(r_0))$, so that $r_0(t) = 0$ for all $t > t_0$. First, let us prove the limit theorem for the maximum of the process $X_0(t)$. Let $T, S \to \infty$, $n \to \infty$, n is an integer, with $S = T/n \to \infty$. We specify their exact limit behavior later. Let us consider the intervals

$$S_k = [kS, (k+1)S - t_0], \quad U_k = [(k+1)S - t_0, (k+1)S], \quad k = 0, 1, \ldots, n-1$$

and their unions,

$$\mathbf{S} = \bigcup_k S_k, \quad \mathbf{U} = \bigcup_k U_k.$$

We have,

$$P\left(\max_{t \in [0,T]} X_0(t) \leqslant u\right) = P\left(\max_{t \in \mathbf{S}} X_0(t) \leqslant u\right)$$

$$- P\left(\max_{t \in \mathbf{S}} X_0(t) \leqslant u, \max_{t \in \mathbf{U}} X_0(t) > u\right), \tag{19.3}$$

and since r_0 has finite support,

$$P\left(\max_{t \in \mathbf{S}} X_0(t) \leqslant u\right) = \left(1 - P\left(\max_{t \in [0, S - t_0]} X_0(t) > u\right)\right)^n.$$

By Theorem 19.1,

$$P\left(\max_{t \in [0, S - t_0]} X_0(t) > u\right) = \frac{S - t_0}{2\pi} e^{-u^2/2} + \Psi(u)$$

$$+ O\left((S - t_0)^2 S^2 e^{-\frac{1}{2}u^2(1+\Delta)}\right)$$

$$= \frac{S}{2\pi} e^{-u^2/2} - \frac{t_0}{2\pi} e^{-u^2/2} + \Psi(u)$$

$$+ O\left(S^2 e^{-\frac{1}{2}u^2(1+\Delta)}\right),$$

where $O(\cdot)$ is uniform in T and u. Since $\Psi(u) = O\left(e^{-u^2/2}\right)$, we continue this as

$$= \frac{S}{2\pi} e^{-u^2/2}\left(1 + O\left(Se^{-\Delta u^2/2}\right) + O\left(S^{-1}\right)\right).$$

Let us denote $l_T = \sqrt{2\ln(T/2\pi)}$ and take $u = l_T + x/l_T$. Then

$$P\left(\max_{t \in [0, S - t_0]} X_0(t) > u\right) = \frac{1}{n} \exp\left(-x - \frac{x^2}{2l_T^2}\right)$$

$$\times \left(1 + O\left(Se^{-\Delta u^2/2}\right) + O\left(S^{-1}\right)\right).$$

Now let $n = \lfloor T^{1-\Delta/2} \rfloor$ ($\lfloor \cdot \rfloor$ is the integer part of a number), and denote

$$p_0(S - t_0) := P\left(\max_{[0,S-t_0]} X_0(t) > u\right).$$

We have,

$$(1 - p_0(S - t_0))^n = \exp\left(n \ln\left(1 - p_0(S - t_0)\right)\right)$$

$$= \exp\left(-n\left(p_0(S - t_0) + O\left(p_0^2(S - t_0)\right)\right)\right). \quad (19.4)$$

Furthermore, by Theorem 19.1,

$$np_0(S - t_0) = \frac{T - nt_0}{2\pi} \exp\left(-\frac{1}{2}(l_T + x/l_T)^2\right) \quad (19.5)$$

$$+ n\left(1 - \Phi(l_T + x/l_T)\right)$$

$$+ O\left(nS^2 \exp\left(-\frac{(1 + \Delta)(l_T + x/l_T)^2}{2}\right)\right).$$

Notice first of all that

$$\frac{T}{2\pi} \exp\left(-\frac{1}{2}(l_T + x/l_T)^2\right) = \exp\left(-x - x^2/2l_T^2\right).$$

Let us show that it is equal to (19.5) up to a negative power of T. If $x \geq -l_T^{3/2}$, then for any $\varepsilon > 0$ and all sufficiently large T,

$$l_T + x/l_T \geq l_T - l_T^{1/2} \geq (1 - \varepsilon)l_T,$$

and, using the above choice of n and properties of the Gaussian tail, we get for some $\delta > 0$, and a sufficiently small ε,

$$n\left(1 - \Phi(l_T + x/l_T)\right) = O(T^{1 - \Delta/2}T^{-1+\varepsilon}) = O(T^{-\delta}),$$

and

$$nS^2 \exp\left(-\frac{(1 + \Delta)(l_T + x/l_T)^2}{2}\right) \leq 2T^{1+\Delta/2} \exp\left(-\frac{(1 + \Delta)(1 - \varepsilon)l_T^2}{2}\right)$$

$$= 2T^{1+\Delta/2}\,(T/2\pi)^{-(1+\Delta)(1-\varepsilon)}$$

$$= O(T^{-\delta}).$$

Hence it follows from (19.5) that

$$\exp\left(-nP\left(\max_{[0,T/n-t_0]} X_0(t) > u\right)\right) = \exp\left(-e^{-x - x^2/2l_T^2}\left(1 + O(T^{-\delta})\right)\right)$$

$$= \exp\left(-e^{-x - x^2/2l_T^2}\right) + O(T^{-\delta'})$$

as $T \to \infty$, for any positive $\delta' < \delta$ and all $x \geq -l_T^{3/2}$. For the second multiplier on the right-hand side of (19.4), from the estimates above,

$$\exp\left(O\left(np_0^2(S-t_0)\right)\right)$$
$$= \exp\left(n^2 p_0^2(S-t_0)O\left(\frac{1}{n}\right)\right)$$
$$= \exp\left(\left(e^{-2x-x^2/l_T^2} + O(T^{-\delta'})\right)O\left(T^{-1+\Delta/2}\right)\right)$$
$$= 1 + O(T^{-\delta''})$$

for any $\delta'' < 1 - \Delta/2$. For the second summand on the right-hand side of (19.3), if $x \geqslant -l_T^{3/2}$, we have,

$$P\left(\max_{t\in S} X_0(t) \leqslant u, \max_{t\in U} X_0(t) > u\right) \leqslant P\left(\max_{t\in U} X_0(t) > u\right)$$
$$\leqslant nP\left(\max_{t\in[0,t_0]} X_0(t) > u\right)$$
$$= O\left(ne^{-u^2/2}\right) = O(1/S) = O(T^{-\Delta/2}).$$

For $x < -l_T^{3/2}$ we have,

$$P\left(\max_{t\in[0,T]} X_0(t) \leqslant l_T + x/l_T\right) \leqslant P\left(\max_{t\in[0,T]} X(t) \leqslant l_T - \sqrt{l_T}\right)$$
$$\leqslant P\left(\max_{t\in S} X(t) \leqslant l_T - \sqrt{l_T}\right)$$
$$= \left(1 - P\left(\max_{[0,S-t_0]} X(t) > l_T - \sqrt{l_T}\right)\right)^n$$
$$\leqslant \exp\left(-Ce^{l_T^{3/2}-l_T/2}\right) = O(T^{-K}), \qquad (19.6)$$

for any $K > 0$.

Thus we have shown that for the family of functions

$$A_T(x) = \begin{cases} e^{-e^{-x-x^2/2l_T^2}}, & x \geqslant -l_T^{3/2}, \\ 0, & x < -l_T^{3/2}, \end{cases}$$

$T > 0$ and $\delta > 0$, uniformly in x,

$$P\left(\max_{t\in[0,T]} X_0(t) \leqslant l_T + x/l_T\right) - A_T(x) = O(T^{-\delta}).$$

Here we used the estimate

$$P\left(\max_{t\in[0,T]} X_0(t) \leqslant 0\right) = O(T^{-K}),$$

which is valid for any $K > 0$. It can be easily derived in a way similar to the derivation of (19.6).

Now let us try to relax the assumption that the covariance function has bounded support, and assume that for the covariance function $r(t)$ of a Gaussian stationary process $X(t)$, along with the conditions of Theorem 18.6, the following holds,

$$\int_0^\infty |r(t)|^a dt < \infty \tag{19.7}$$

with some positive a. Using this, let us estimate the right-hand side of (18.15) for $X_1 = X$. Let us denote it by Δ.

First, however, let us prove that (19.7) implies that $r(t) \to 0$ as $t \to \infty$. Indeed, if for some t_0, $r(t_0) \geq \varepsilon > 0$ then, since $|r(t_0) - r(t_0 + s)| \leq \sqrt{2(1 - r^2(s))}$ (a property of non-negative definite functions), by (19.1)–(19.2), we have that $|r(t_0) - r(t_0 + s)| \leq 3|s|$ for $|s| \leq \delta$ with some $\delta > 0$ not depending on t_0. Hence $r(t_0 + s) \geq r(t_0) - 3|s| \geq \varepsilon - 3|s|$. It means that one can place a triangle with the base equal to at least $\min(3\delta, \varepsilon)$ and the height equal to at least $\min(\varepsilon, \delta/3)$ under the curve $y = r(t)$. The area of this triangle will be no less than $\frac{1}{9}\min(\varepsilon^2, \delta^2)$. Now, assuming that $r(t)$ does not tend to zero, one can find infinitely many such triangles and, thus, the integral diverges. The case $r(t_0) < -\varepsilon$ can be handled in the same way. Therefore, for some T_0 and sufficiently small $\varepsilon > 0$, $|r(t)| \leq \varepsilon$ as $t \geq T_0$. Hence

$$\Delta \leq C\left[Te^{-\frac{u^2}{2(1+\delta)}} + TT_0 e^{-\frac{u^2}{1+\rho}} + T\int_{T_0}^T |r(t)| e^{-\frac{u^2}{1+|r(t)|}} dt\right] \tag{19.8}$$

(taking $\delta_1 = \min\left(\delta, \frac{1-\rho}{2(1+\rho)}\right)$, $\rho := \max_{[t_0, T_0]} |r(t)| < 1$)

$$\leq C_1\left[Te^{-\frac{1}{2}u^2(1+\delta_1)} + Te^{-\frac{u^2}{1+\varepsilon}}\int_{T_0}^T |r(t)| dt\right]$$

(by Minkowski's inequality with $a' = \max(1, a)$)

$$\leq C_1\left[Te^{-\frac{1}{2}u^2(1+\delta_1)} + TT^{1-1/a'}e^{-\frac{u^2}{1+\varepsilon}}\left[\int_{T_0}^T |r(t)|^{a'}\right]^{1/a'} dt\right]$$

$$\leq C_1\left[Te^{-\frac{1}{2}u^2(1+\delta_1)} + T^{2-1/a'}e^{-u^2}e^{-\frac{\varepsilon u^2}{1+\varepsilon}} dt\right]$$

$$= C_1\left[Te^{-\frac{1}{2}u^2}e^{-\frac{1}{2}\delta_1 u^2} + T^2 e^{-u^2}T^{-1/a'}e^{-\frac{\varepsilon u^2}{1+\varepsilon}} dt\right].$$

Now let us choose ε small enough to have

$$\delta_2 := \frac{1}{a'} - \frac{2\varepsilon}{1+\varepsilon} > 0.$$

Then we can conclude that for $u = l_T + x/l_T$ and $x \geq -l_T^{3/2}$, $\Delta \leq C_2 T^{-\delta_3}$ for any $\delta_3 \in (0, \min(\delta_1, \delta_2))$ and a suitable C_2. Moreover we already know from (19.6) that for $x < -l_T^{3/2}$,

$$P\left(\max_{t\in[0,T]} X_0(t) \leqslant l_T + x/l_T\right) \leqslant CT^{-K}.$$

Thus this inequality holds for $X_1 = X$ as well. Furthermore, it is easy to calculate that

$$l_T^2\left(A_T(x) - e^{-e^{-x}}\right) \rightarrow \frac{1}{2}e^{-e^{-x}}e^{-x}x^2 \tag{19.9}$$

as $T \rightarrow \infty$ uniformly in $x \in \mathbb{R}$. Notice that the right-hand side of (19.9) vanishes as $x \rightarrow \pm\infty$. Therefore (19.9) holds with

$$P\left(\max_{t\in[0,T]} X(t) \leqslant l_T + x/l_T\right)$$

in place of $A_T(x)$. Let us formulate what we have proven.

Theorem 19.3. *Let $X(t)$, $t \in \mathbb{R}$, be a twice differentiable in square mean Gaussian stationary process with $EX(t) = 0$, $EX^2(t) = 1$, $EX'(t)^2 = 1$. Assume that*

$$\int_0^T |r(t)|^a dt < \infty$$

holds for its covariance function r and some $a > 0$. Denote $l_T = \sqrt{2\ln\frac{T}{2\pi}}$. Then

1. For some $\gamma > 0$,

$$P\left(\max_{t\in[0,T]} X(t) \leqslant l_T + x/l_T\right) - A_T(x) = O(T^{-\gamma}), \quad T \rightarrow \infty$$

uniformly in $x \in \mathbb{R}$.
2. Also,

$$l_T^2\left(P\left(\max_{t\in[0,T]} X(t) \leqslant l_T + x/l_T\right) - e^{-e^{-x}}\right) \rightarrow \frac{1}{2}e^{-e^{-x}}e^{-x}x^2,$$

as $T \rightarrow \infty$, uniformly in $x \in \mathbb{R}$.

It follows from the second statement that the rate of convergence of the distribution of the maximum to the Gumbel distribution is logarithmic. It also gives the second term of the asymptotic expansion for the probability. The first statement gives the sequence of approximating functions that approaches the Gumbel distribution with the power rate.

Geometry of High Excursions of Stationary Smooth Fields

20.1 Introduction

The present lecture is an expanded version of my talk "Excursions of Gaussian Fields: Comparison Theorems, Geometrical Probabilities and Euler Characteristics" that I gave to the Moscow Mathematical Society on October, 11, 2005. Here I consider topological characteristics of the level sets

$$\{\mathbf{t} \in \mathbb{R}^n \colon X(\mathbf{t}) = u\}$$

for a smooth Gaussian stationary (homogeneous) field in the limit of large u. In dimension one this problem is extremely simple, as "level sets" are just the crossing points of u. In the multi-dimensional case we need to deal with $(n-1)$-dimensional surfaces. The geometry of these surfaces is the subject of our investigation. In contrast to other lectures, the technical background here is highly complex, so we restrict ourselves to a narrative, describing most interesting facts and notions in relatively simple terms. One can find complete proofs and other related facts in Piterbarg [1996]. The monographs Adler [1981] and Adler and Taylor [2007] use a different approach to the same subject.

Let us recall the following result for Gaussian processes with smooth trajectories from Lecture 15 (with obvious generalizations).

Theorem 20.1. *Let $X(t)$, $t \in [0,T]$, be a Gaussian process with zero mean and a.s. twice differentiable trajectories. Assume that for any t_1, t_2, $t_1 \neq t_2$, the covariance matrix of the vector $(X(t_1), X(t_2), X'(t_1), X'(t_2))$ is nondegenerate, and $\mathrm{E}X^2(t) \equiv \mathrm{E}X'(t)^2 \equiv 1$. Then there exist C and $\delta > 0$ such that*

$$\mathrm{P}(\max_{t\in[0,T]} X(t) > u) = \frac{T}{2\pi} e^{-u^2/2} + 1 - \Phi(u) + R(u) \tag{20.1}$$

with

$$0 \leqslant R(u) \leqslant CT^2 \exp\left(-\frac{1}{2}(1+\delta)u^2\right). \tag{20.2}$$

We would like to generalize this result to Gaussian fields. But first we need to formulate appropriate conditions on the fields and the parametric sets where the fields are defined.

20.2 Random Fields. Definitions and Assumptions

Let us consider a Gaussian random field $X(\mathbf{t})$, $\mathbf{t} \in T \subset \mathbb{R}^n$, that is, $X(\mathbf{t}) = X(\mathbf{t},\omega) : \mathbb{R}^n \times \Omega \to \mathbb{R}$, where $(\Omega,\mathcal{F},\mathrm{P})$ is the basic probability space and for any $(\mathbf{t}_1,\dots,\mathbf{t}_N)$, the distribution of the vector $(X(\mathbf{t}_1),\dots,X(\mathbf{t}_N))$ is Gaussian with parameters $m(\mathbf{t}_i), i = 1,\dots,N$, $r(\mathbf{t}_i,\mathbf{t}_j), i,j = 1,\dots,N$.

20.2.1 Smoothness and Non-Degeneracy

We assume that the covariance function $r(\mathbf{s},\mathbf{t})$ is six times differentiable, so that the field X is three times differentiable in the square mean sense. We can consider a version of the field with a.s. twice continuously differentiable trajectories[1].

Condition 20.2. *Let us denote*

$$\mathbf{Z}(\mathbf{t}) := \left(X(\mathbf{t}), \nabla X(\mathbf{t}), \frac{\partial^2}{\partial t_i \partial t_j} X(\mathbf{t}), i \geq j, \frac{\partial^3}{\partial t_i \partial t_j \partial t_k} X(\mathbf{t}), i \geq j \geq k \right).$$

We assume that for any $\mathbf{s} \neq \mathbf{t}$ the distribution of the vector $(\mathbf{Z}(\mathbf{s}), \mathbf{Z}(\mathbf{t}))$ is non-degenerate.

20.2.2 Parametric Set

We assume that the parametric set $T \subset \mathbb{R}^n$ is a *simple cell complex* of the class $C^3(M,K,\psi,\varepsilon)$, $K,M,\varepsilon > 0, \psi \in (0,1)$. Later we will give the exact definition. Some examples of such sets include

- Finitely connected sets with non-degenerate smooth boundaries (a ball, a torus, a sphere);
- Intersections of finite numbers of such sets with some non-degeneracy conditions at all points of intersections of their boundaries;
- Closed convex polyhedrons with finite number of vertices, as well as their sufficiently smooth non-degenerate transforms.

Let us now state the general definition.

Definition 20.3. *The couple $(\mathcal{U},\alpha)$, $\mathcal{U} \subset \mathbb{R}^n$, $\dim \mathcal{U} = l, l \leqslant n$, $\alpha : \mathcal{U} \to \mathbb{R}^l$ a diffeomorphism, is called a chart of the class C^ν, if*

---

[1] The reader should carefully verify these statements by following the arguments of Lecture 7 and Introduction to Lecture 8

1. $\alpha \in C^\nu$;
2. The set $\alpha \mathcal{U} \subset \mathbb{R}^l$ is open and bounded;
3. There exists the inverse map $\alpha^{-1} : \alpha \mathcal{U} \to \mathbb{R}^l$, that also belongs to C^ν.

A one-point set $\{\mathbf{x}\} \in \mathbb{R}^n$ paired with an arbitrary α of class C^ν, that is, $(\{\mathbf{x}\}, \alpha)$, is also considered to be a chart of the class C^ν.

Definition 20.4. *We say that the chart of class C^ν belongs to the class (C^ν, M), $M > 0$, if all partial derivatives of α and α^{-1} up to the order ν are uniformly bounded by M. We also say that such α belongs to $C^\nu(M)$.*

Definition 20.5. *A closed set $M \subset \mathbb{R}^n$ is called a simple cell complex of class C^ν if there exists a finite collection of charts $(\mathcal{M}_i, \alpha_i)$, $i = 1, \ldots, N$, $\mathcal{M}_i \subset \mathbb{R}^n$, all are of class C^ν, $\mathcal{M}_i$ are pairwise disjoint and $M = \bigcup_{i=1}^N \mathcal{M}_i$. Any such collection is called the stratification of M and is denoted by $s(M)$.*

Definition 20.6. *Let $\Pi \subset \mathbb{R}^n$ be a polyhedral angle, that is an intersection of linear subspaces,*

$$\Pi = \bigcap \{\mathbf{x} : (\mathbf{a}_i, \mathbf{x}) \geqslant 0\}, \quad \mathbf{a}_i \in \mathbb{R}^n.$$

We say that Π is a ψ-angle, $0 < \psi < 1$, if for any k, l either

$$|(\mathbf{a}_k, \mathbf{a}_l)| \leqslant \psi |\mathbf{a}_k| |\mathbf{a}_l|$$

or

$$|(\mathbf{a}_k, \mathbf{a}_l)| = |\mathbf{a}_k| |\mathbf{a}_l|.$$

In other words, either both vectors are collinear, or the absolute value of the cosine of the angle between them is not greater than ψ.

Notice that the dimension of a ψ-angle can be less than n.

Definition 20.7. *We say that the simple cell complex $S \subset \mathbb{R}^n$ is of the class $C^3(M, K, \psi, \varepsilon)$, $K, M, \varepsilon > 0, \psi \in (0,1)$, if*

1. *There exists a stratification $s(S)$ such that $\mathrm{Card}(s(S)) \leqslant K$, and for all $(\mathcal{U}, \alpha) \in s(S)$, $(\mathcal{U}, \alpha) \in (C^3, M)$.*
2. *For any $\mathbf{s}, \mathbf{t} \in S$ with $|\mathbf{s} - \mathbf{t}| \leqslant \varepsilon$ there exists a diffeomorphism $\beta \in C^3(M)$ of the ball $B_{2\varepsilon}(\mathbf{t}) = \{\mathbf{v} : |\mathbf{v} - \mathbf{t}| \leqslant 2\varepsilon\}$, a neighbourhood V of the segment $\{h\beta(\mathbf{s}) + (1 - h)\beta(\mathbf{t}), 0 \leqslant h \leqslant 1\}$ and a ψ-angle Π such that*

$$\beta(B_{2\varepsilon}(\mathbf{t}) \cap S) \cap V = \Pi \cap V.$$

Roughly, all the "angles" in S are ψ-angles.

20.3 Comparison Method

While we used the moment method to prove Theorem 20.1 previously, here we will apply the comparison method, a multi-dimensional analog to the (one-dimensional) case considered in Lecture 19. We state the following theorem without proof.

Theorem 20.8. *Let the parametric set T and two Gaussian fields $X_0(t)$ and $X_1(t)$, $t \in T$, with zero means satisfy the assumptions above. Assume also that $EX_0(t)^2 \equiv EX_1(t)^2$ and $\mathrm{Cov}\nabla X_0(t) \equiv \mathrm{Cov}\nabla X_1(t)$. Then there exist $L, \delta > 0$ such that for all u,*

$$\left| P(\max_{t \in T} X_1(t) \leqslant u) - P(\max_{t \in T} X_0(t) \leqslant u) \right| \leqslant L \exp\left(-\frac{1}{2}(1+\delta)u^2\right). \quad (20.3)$$

The proof of Theorem 20.8 is somewhat similar to the proofs of Theorems 18.2 and 18.3. It starts with the discrete-time comparison theorem, Theorem 18.1, applied to an appropriate grid in T. However, in multiple dimensions even this step is non-trivial since the boundary of T consists of cells of different dimensions. Indeed, we introduced all these differential topology type conditions and assumptions above precisely so that we can define suitable grids and pass to a limit eventually.

The following theorem is a generalization of Theorem 18.6. It is not used in these lectures but is useful for studying large excursions of Gaussian homogeneous fields ($r(s,t) \equiv r(t-s)$ and $EX(t) \equiv \mathrm{Const}$) on unboundedly dilative sets.

Theorem 20.9. *Let, in addition to all the conditions of Theorem 20.8, the fields X_i, $i = 1,2$ be homogeneous with $r_i(t) \to 0$ as $t \to \infty$, $i = 0,1$. Then L and $\delta > 0$ can be chosen not depending of T and such that for some t_0,*

$$\left| P(\max_{t \in T} X_1(t) \leqslant u) - P(\max_{t \in T} X_0(t) \leqslant u) \right| \leqslant L \exp\left(-\frac{1}{2}(1+\delta)u^2\right)$$

$$+ \int_{T \times T \cap \{|t-s| > t_0\}} |r_1(t-s) - r_0(t-s)| \exp\left(-\frac{u^2}{1+\widehat{r}(t-s)}\right) ds dt,$$

where $\widehat{r}(t) = \max(r_0(t), r_1(t))$.

We shall apply Theorem 20.8 to derive an asymptotic expansion of the distribution of the maximum of the Gaussian random field, as $u \to \infty$, up to the order of the right-hand side of (20.3). In order to do this, we need a standard Gaussian field for which one can approximate the maximum distribution up to the required order. In the one-dimensional case we would use the Gaussian sinusoid, $X \cos t + Y \sin t$, with independent standard Gaussian X, Y. We are going to do something similar in the multi-dimensional case.

20.4 Choice of the Standard Field

The field

$$Y_n(t) = n^{-1/2} \sum_{i=1}^{n} (X_i \cos t_i + Y_i \sin t_i),$$

$t = (t_1,\ldots,t_n)$, with $(X_i, Y_i, i = 1,\ldots,n)$, independent Gaussian standard variables, is a Gaussian homogeneous field. Let us find the distribution of its maximum on the multidimensional rectangle $K(\mathbf{T}) := [0,T_1] \times \cdots \times [0,T_n]$ for $T_i \in [0,\pi)$, $i = 1,\ldots,n$. The Gaussian stationary process $Y(t) = X_1 \cos t + Y_1 \sin t$, $t \in [0,T]$, $T < \pi$, can be written as $Z(t) = \sqrt{X_1^2 + Y_1^2} \cos(t - \varphi)$, where the random variable φ is uniformly distributed on $[0,2\pi]$ and does not depend of $\sqrt{X_1^2 + Y_1^2}$. Since $T < \pi$, the two-dimensional distributions of Z are non-degenerate, and nor are the joint distribution of the process and its derivative at the same point. Hence the mean number of up-crossings of u can be calculated in the standard way,

$$EN_+(T) = \frac{T}{2\pi} e^{-u^2/2}.$$

Furthermore, let us write

$$P(\max_{[0,T]} Y(t) > u) = P(Y(0) > u) + P(Y(0) \leqslant u, N_+(T) \geqslant 1),$$

and notice that for $u > 0$ the set $\{Y(0) > u, N_+(T) \geqslant 1\}$ is empty since there is not enough time for a sinusoid to go first under u and then go beyond u till T. Hence,

$$P(\max_{[0,T]} Y(t) > u) = P(Y(0) > u) + P(N_+(T) \geqslant 1).$$

We have that $N_+ = 0$ or $N_+ = 1$, that is, the last summand is the mean of the number of up-crossings. Thus

$$P(\max_{[0,T]} Y(t) > u) = \frac{T}{2\pi} e^{-u^2/2} + 1 - \Phi(u).$$

Let us now turn to Y_n. Denoting $\eta_k = \max_{[0,T_k]}(X_k \cos t_k + Y_k \sin t_k)$, we come to a problem of calculating the probability

$$P_u = P(\eta_1 + \cdots + \eta_n > \sqrt{n}u).$$

The probability density of η_k at points $x > 0$ is equal to $f(x) = \varphi(x) - \frac{T_k}{\sqrt{2\pi}}\varphi'(x)$. In order to evaluate the asymptotic behavior of the probability, we separate the negative values of the argument,

$$P_u = \int_{y_1 + \cdots + y_n \geqslant \sqrt{n}u} \prod_{k=1}^{n} f(y_k) dy_1 \ldots dy_k$$

$$= \int_{\substack{y_1 + \cdots + y_n \geqslant \sqrt{n}u \\ y_i > 0, i=1,\ldots,n}} \prod_{k=1}^{n} f(y_k) dy_1 \ldots dy_k$$

$$+ \int_{\substack{y_1 + \cdots + y_n \geqslant \sqrt{n}u \\ \exists i: y_i < 0}} \prod_{k=1}^{n} f(y_k) dy_1 \ldots dy_k. \tag{20.4}$$

The second integral on the right-hand side does not exceed the sum of probabilities

$$\sum_{i=1}^{n} P(\eta_1 + \cdots + \eta_n - \eta_i > \sqrt{n}u). \tag{20.5}$$

The first one is equal to

$$\int_{y_1 + \cdots + y_n \geqslant \sqrt{n}u} \prod_{k=1}^{n} \left(\varphi(y_k) - \frac{T_k}{\sqrt{2\pi}} \varphi'(y_k) \right) dy_1 \ldots dy_k$$

$$- \int_{\substack{y_1 + \cdots + y_n \geqslant \sqrt{n}u \\ \exists i: y_i < 0}} \prod_{k=1}^{n} \left(\varphi(y_k) - \frac{T_k}{\sqrt{2\pi}} \varphi'(y_k) \right) dy_1 \ldots dy_k. \tag{20.6}$$

Note that now the integrands are not densities! We have for all x and $T_k < \pi$,

$$\left| \varphi(x) - \frac{T_k}{\sqrt{2\pi}} \varphi'(x) \right| = \varphi(x) \left| 1 + \frac{T_k x}{\sqrt{2\pi}} \right|$$

$$\leqslant \frac{1}{\sqrt{2\pi}} (1 + \sqrt{2 + n^2}) \exp \left(-\frac{1}{2} x^2 \left(1 - \frac{1}{n^2} \right) \right),$$

therefore the second integral on the right-hand side of (20.6) is at most

$$\left(1 + \sqrt{2 + n^2}\right)^n \int_{\substack{y_1 + \cdots + y_n \geqslant \sqrt{n}u \\ \exists i: y_i < 0}} \prod_{k=1}^{n} \varphi\left(y_i \sqrt{1 - n^{-2}}\right)$$

$$= \left(1 + \sqrt{2 + n^2}\right)^n P\left(X_1 + \cdots + X_n \geqslant u\sqrt{n}\sqrt{1 - n^{-2}}, \exists i : X_i < 0\right)$$

$$\leqslant \left(1 + \sqrt{2 + n^2}\right)^n nP\left(X_1 + \cdots + X_{n-1} \geqslant u\sqrt{n}\sqrt{1 - n^{-2}}\right),$$

with X_i being independent standard Normal variables. In turn, using the standard bound for the tail of the Normal distribution, we see that the latter probability is at most

$$\frac{1}{\sqrt{2\pi}u\sqrt{n}\sqrt{1 - n^{-2}}} \exp \left(\frac{-nu^2(1 - n^{-2})}{2n - 2} \right).$$

Let us now write the first integral on the right-hand side of (20.6) as an integral of the convolution,

$$\int_y^\infty \left(n^{n/2} *_{i=1}^n \left((y/\sqrt{n}) - \frac{T_i}{\sqrt{2\pi}} \varphi'(y/\sqrt{n}) \right) \right) dy. \tag{20.7}$$

The Fourier transform of the integrand is equal to

$$\prod_{k=1}^n \left(e^{-t^2/2n} - \frac{T_k it}{\sqrt{2\pi n}} e^{-t^2/2n} \right)$$

$$= e^{-t^2/2} \left(1 - \frac{\sum T_k}{\sqrt{2\pi n}} it + \frac{\sum T_k T_l}{\left(\sqrt{2\pi n}\right)^2} (it)^2 - \cdots + (-1)^n \frac{\prod T_k}{\left(\sqrt{2\pi n}\right)^n} \right).$$

Denote by $\Sigma_\nu(\mathbf{T})$ the standard homogeneous polynomial of the components of $\mathbf{T} = (T_1, \ldots, T_n)$ of degree ν. Applying the inverse Fourier transform, we get that (20.7) is equal to

$$\int_u^\infty \left(\varphi(x) - \frac{\Sigma_1(\mathbf{T})}{\sqrt{2\pi n}} \varphi'(x) + \frac{\Sigma_2(\mathbf{T})}{(\sqrt{2\pi n})^2} \varphi''(x) + \cdots + \frac{\Sigma_n(\mathbf{T})}{(\sqrt{2\pi n})^n} \varphi^{(n)}(x) \right) dx.$$

Now, applying all the above estimates we get that the first integral on the right-hand side of (20.6) is equal to

$$1 - \Phi(u) + \frac{\Sigma_1(\mathbf{T})}{(2\pi n)^{1/2}} \varphi'(u) + \cdots + \frac{\Sigma_n(\mathbf{T})}{(2\pi n)^{n/2}} \varphi^{(n)}(u)$$

$$+ O\left(\exp\left(-\frac{1}{2} u^2 (1 + 1/n) \right) \right), \quad u \to \infty. \tag{20.8}$$

To estimate the second integral on the right-hand side of (20.6) we consider its bound (20.5). Under the probability signs we have the sum of $(n-1)$ standard Normal variables. Writing

$$\sqrt{n}u = \sqrt{n-1}u\sqrt{1 + 1/(n-1)},$$

we obtain an estimate of the difference of integrals in (20.6) for the level

$$u_1 = u\sqrt{1 + 1/(n-1)},$$

which gives the order less than $\exp((1+1/n)u^2/2)$. Repeating this procedure $n-1$ times we get that P_u is also equal to (20.8) as $u \to \infty$.

In order to generalize this result to arbitrary parametric sets we need some definitions and facts from integral geometry. Let A be a convex compact in $\mathbb{R}^n$, $B_\rho = \{\mathbf{x} : \|\mathbf{x}\| \leqslant \rho\}$ be the ball of radius ρ and $A \oplus B_\rho$ be their Minkowski sum. By Steiner's formula we have for its Euclidean volume,

$$V(A \oplus B_\rho) = \sum_{j=0}^n \binom{n}{j} \rho^j W_j(A),$$

where $W_j(A)$ are the Minkowski functionals. This formula can be an intuitive definition of the functionals. For example, the zero functional is the volume, the one-functional is proportional to the $(n-1)$-dimensional surface volume (one can get this subtracting W_0 and dividing both sides by ρ, and setting ρ to zero); the two-functional is proportional to the $(n-2)$-dimensional volume of all "edges", and so on. This also can be obtained by suitable limit passages as $\rho \to 0$. One can find exact derivations and lots of other interesting stuff about integral geometry and geometric probabilities in Matheron [1976] and Santalo [2004].

Now let us come back to P_u and recall the expressions for Hermite polynomials,

$$H_v(u) = \varphi^{(n)}(u)/\varphi(u).$$

It is very easy to obtain the expressions for Minkowski functionals of multidimensional rectangles,

$$W_v(K(\mathbf{T})) = \frac{\omega_v}{\binom{n}{v}} \Sigma_{n-v}(\mathbf{T}),$$

with ω_v the volume of the unit ball in $\mathbb{R}^v$. Thus we can rewrite P_u as

$$P_u = \varphi(u) \sum_{v=0}^{n-1} \frac{\binom{n}{v} H_{n-1-v}(u)}{\omega_v (2\pi n)^{(n-v)/2}} W_v(K(\mathbf{T})) + \int_u^\infty \varphi(x)dx$$
$$+ O\left(\exp\left(-\frac{1}{2}u^2(1 + 1/n)\right)\right), \quad u \to \infty. \tag{20.9}$$

20.5 Extension to an Arbitrary Field on a Small Rectangle

Lemma 20.10. *Let a Gaussian field $X(\mathbf{t})$ with zero mean, unit variance and the covariance matrix of its gradient*

$$\Lambda_2 = \mathbb{E}\nabla X(\mathbf{t})\nabla X(\mathbf{t})^\top = n^{-1}I,$$

where I is the unit matrix, satisfy the non-degeneracy Condition 20.2 on a multi-dimensional rectangle $K(\mathbf{S})$, $\mathbf{S} = (S_1,\dots,S_n)$, $0 < S_i < \pi$, $i = 1,\dots,n$. Then there exist $\rho > 0$ and Q_i, $0 < Q_i < S_i$, $i = 1,\dots,n$, such that for any $\mathbf{T} = (T_1,\dots,T_n)$, $0 \leqslant T_i \leqslant Q_i$, $i = 1,\dots,n$,

$$\mathrm{P}(\max_{\mathbf{t}\in K(\mathbf{T})} X(\mathbf{t}) > u) = p(T_1,\dots,T_n,u) + O(\exp\left(-u^2(1 + \rho)/2\right),$$

as $u \to \infty$ where $p(T_1,\dots,T_n,u)$ is the right-hand side of (20.9).

Proof. We cannot apply Theorem 20.8 directly to $X(\mathbf{t})$ and $Y_n(\mathbf{t})$, because it is impossible to fit the covariance function of X at zero by re-scaling the covariance function of Y_n, and because of degeneracy of the required finite-dimensional distributions of Y_n. Nevertheless, the expansion (20.9) and Theorem 20.8 play a crucial role in the proof. Consider the cosine field

$$Y_N(t\sqrt{N/n}), \quad t \in \mathbb{R}^N, \quad N \geqslant n,$$

and its restriction $Y_N^n(t)$ on $\mathbb{R}^n = \{t = (t_1,\ldots,t_n,0,\ldots,0)\}$. Its covariance function is equal to

$$r_N^n(t) = \frac{N - n + \sum_{i=1}^n \cos\left(t_i\sqrt{N/n}\right)}{N}$$

$$= 1 - \frac{1}{2n}|t|^2 + \frac{N}{4!n^2}\sum_{i=1}^n t_i^4 + O(|t|^6), \quad |t| \to 0.$$

Furthermore, consider a Gaussian homogeneous field $X_0(t)$ with zero mean and the covariance function $r_0(t) = \exp(-|t|^2/2n)$. It is clear that

$$r_0(t) = 1 - \frac{1}{2n}|t|^2 + \frac{1}{8n^2}|t|^4 + O(|t|^6), \quad |t| \to 0.$$

It follows from this and previous estimates that for a sufficiently large $N > 6$, there exists a neighbourhood U_1 of zero such that $r_N^n(t) \geqslant r_0(t)$, $t \in U_1$. Take $M > N$ and $0 < \varepsilon < (M-6)/(M-N)$. One can find a neighbourhood U_2 of zero such that $r_1(t) \geqslant r_N^n(t)$, $t \in U_2$, where $r_1(t) = (1-\varepsilon)r_M^n(t) + \varepsilon r_0(t)$. One can make sure of this by summing the correspondingly weighted expansions of both covariance functions. Thus we have in $U = U_1 \cap U_2$,

$$r_1(t) \geqslant r_N^n(t) \geqslant r_0(t).$$

Let $X_1(t)$ be a Gaussian homogeneous field with zero mean and the covariance function $r_1(t)$. By Slepian's theorem, for any closed $A \subset U$ and all u,

$$P(\max_{t \in A} X_1(t) > u) \leqslant P(\max_{t \in A} Y_N^n(t) > u) \leqslant P(\max_{t \in A} X_0(t) > u).$$

The fields $X_0(t)$ and $X_1(t)$ satisfy all the conditions of Theorem 20.8. All closed rectangles lying in U are obviously simple cell complexes of the class $C^3(M,K,\psi,\varepsilon)$, for the same M,K,ψ and ε. Moreover,

$$P\left(\max_{0 \leqslant t_i \leqslant T_i, i=1,\ldots,n} Y_N^n(t) > u\right) = P\left(\max_{0 \leqslant t_i \leqslant T_i\sqrt{N/n}, i=1,\ldots,n} Y_N(t) > u\right),$$

where $T_{n=1} = \cdots = T_N = 0$. The statement of the lemma now follows for the field $X_0(t)$, provided $T_i < \pi\sqrt{n/N}$ for all i and using the fact that Minkowski functionals do not depend on shifts. The statement of the lemma follows now from Theorem 20.8.

20.6 Extension to an Arbitrary Parametric Set

Denote by G_n the factor-group of the isometry group in $\mathbb{R}^n$ (the group of distance-preserving maps) with respect to all translations in $\mathbb{R}^n$. Thus we write $gA \subset B, g \in G_n$, if there exists a translation l such that $lg_0 A \subset B$, where g_0 is a rotation around zero. We need the following corollary to Theorem 20.8.

Corollary 20.11. *Let $X(\mathbf{t}), \mathbf{t} \in T$ be a Gaussian field with zero mean, unit variance and the covariance matrix of its gradient $\Lambda_2 \equiv n^{-1}I$. Assume the field is non-degenerate in the sense of Condition 20.2. Let $S \subset T$ be a simple cell complex of the class $C^3(M,K,\psi,\varepsilon)$. Then one can find $L,\rho > 0$ such that for any $g \in G_n$ with $gS \subset T$,*

$$\left| P(\max_{\mathbf{t} \in S} X(\mathbf{t}) > u) - P(\max_{\mathbf{t} \in gS} X(\mathbf{t}) > u) \right| \leqslant L\exp(-u^2(1+\rho)/2).$$

Proof follows from the observation that the pair $X(\mathbf{t})$ and $X(g^{-1}\mathbf{t})$ satisfies the conditions of Theorem 20.8.

Let us write down the cosine field $Y_N(\mathbf{t}), \mathbf{t} \in \mathbb{R}^N$, as

$$Y_N(\mathbf{t}) = N^{-1/2} \sum_{k=1}^{N} Z_k \cos(t_k - \phi_k),$$

where $\boldsymbol{\phi} = (\phi_1, \ldots, \phi_N) \in [0, 2\pi)^N$ is a unique point of maximum of Y_N on this cube. The random variables $Z_1, \ldots, Z_N, \phi_1, \ldots, \phi_N$ are independent and the vector $\boldsymbol{\phi}$ is uniformly distributed in $[0, 2\pi)^N$. For a set A, let us denote

$$A_{\mathrm{mod}2\pi} = \bigcup_{\mathbf{m} \in \mathbb{Z}^n} (A + 2\pi\mathbf{m}).$$

Lemma 20.12. *The set*

$$B_u^1 = \{\mathbf{t} : Y_N(\mathbf{t}) \geqslant u\} \cap \left\{ \boldsymbol{\phi} + [-\pi/2, \pi/2]^N \right\}_{\mathrm{mod}\,2\pi} \cap [0,\pi)^N$$

is a.s. convex in $\mathbb{R}^N$.

Proof. The matrix of the second derivatives of $Y_N(\mathbf{t})$ on $\left\{ \boldsymbol{\phi} + [-\pi/2, \pi/2]^N \right\}_{\mathrm{mod}\,2\pi}$ is a.s. negative-definite. Indeed, for all i,j, with probability one,

$$\frac{\partial}{\partial t_i \partial t_j} Y_N(\mathbf{t}) = 0, \quad i \neq j,$$

$$\frac{\partial^2}{\partial t_i^2} Y_N(\mathbf{t}) = -Z_i \cos(t_i - \phi_i) < 0.$$

Furthermore, the set $\left\{\boldsymbol{\phi} + [-\pi/2,\pi/2]^N\right\}_{\mathrm{mod}2\pi} \cap [0,\pi)^N$ is connected for any $\boldsymbol{\phi}$ hence it is convex. Thus B_u^1 is an intersection in $\mathbb{R}^{N+1} = \{y,t_1,\ldots,t_N\}$ of two convex sets, $\{y \geq u\}$ and

$$\left\{\mathbf{t} : \boldsymbol{\phi} + \left\{\mathbf{t} : N^{-1/2}\sum_{k=1}^{N} Z_k \cos t_k \geq u\right\}\right\} \cap \left\{\boldsymbol{\phi} + [-\pi/2,\pi/2]^N\right\}_{\mathrm{mod}2\pi} \cap [0,\pi)^N.$$

Lemma 20.13. *There exist* $L,\rho_1 > 0$ *such that for any Borel* $A \subset [0,\pi)^N$,

$$\left|P(\sup_A Y_N(\mathbf{t}) > u) - P(A \cap B_u^1 \neq \varnothing)\right| \leq L\exp(-u^2(1+\rho_1)/2).$$

Proof. By our construction,

$$\{\sup_A Y_N(\mathbf{t}) > u\} \supset \{A \cap B_u^1 \neq \varnothing\}.$$

Furthermore,

$$\{\sup_A Y_N(\mathbf{t}) > u\}\backslash\{A \cap B_u^1 \neq \varnothing\} = \{\sup_A Y_N(\mathbf{t}) > u\} \cap \{A \cap B_u^1 = \varnothing\}$$

$$= \{\sup_A Y_N(\mathbf{t}) > u\} \cap \left\{A \cap \left\{\boldsymbol{\phi} + [-\pi/2,\pi/2]^N\right\}_{\mathrm{mod}\,2\pi} = \varnothing\right\}$$

$$\subset \left\{\max_\Gamma Y_N(\mathbf{t}) > u\right\} = \bigcup_{i=1}^{2N}\left\{\max_{\gamma_i} Y_N(\mathbf{t}) > u\right\},$$

where $\Gamma = \cup\gamma_i$, γ_i are the sides of the cube $\boldsymbol{\phi} + [-\pi/2,\pi/2]^N$. From here we get for $N > 1$ that

$$P\{\sup_A Y_N(\mathbf{t}) > u\}\backslash\{A \cap B_u^1 \neq \varnothing\}$$

$$\leq 2NP\left(\max_{\substack{-\pi/2\leq t_i\leq\pi/2, i=1,\ldots,N-1,\\ t_N=\pi/2}} Y_N(\boldsymbol{\phi}+\mathbf{t}) > u\right)$$

$$\leq 2NP\left(N^{-1/2}\sum_{i=1}^{N-1} Z_i > u\right)$$

$$= 2NP\left((N-1)^{-1/2}\sum_{i=1}^{N-1} Z_i > u\sqrt{\frac{N}{N-1}}\right)$$

$$\leq 2NP\left(Y_{N-1}(0) > u\sqrt{\frac{N}{N-1}}\right)$$

$$\leq \frac{2N}{u\sqrt{2\pi}}\exp\left(-\frac{u^2}{2}\left(1+\frac{1}{N-1}\right)\right).$$

In the case $N = 1$, this chain should be interrupted on the second step, where the probability becomes equal to zero for $u > 0$.

20.7 General Case

Now we are able to derive the asymptotic estimate similar to (20.1). We need Hadwiger's theorem, Hadwiger [1957]. First let us give a required definition.

Definition 20.14. *A functional ψ on the set of convex compact sets $\mathcal{K}$ is called C-additive if from $K,K' \in \mathcal{K}$ and $K \cup K' \in \mathcal{K}$ it follows that*

$$\psi(K \cap K') + \psi(K \cup K') = \psi(K) + \psi(K')$$

Theorem 20.15 (Hadwiger's theorem). *If C-additive functional ψ on the space of convex compact sets increases with respect to the natural partial order of sets and is invariant with respect to the group of Euclidean transforms (translations and rotations), then there exist non-negative $a_0,\dots,a_n$ such that*

$$\psi = \sum_{j=0}^{n} a_j W_j.$$

Theorem 20.16. *Let $X(\mathbf{t})$, $\mathbf{t} \in T \subset \mathbb{R}^n$, T is an open set, be a Gaussian random field with zero mean, unit variance and the covariance matrix of its gradient identically equal to $\Lambda_2 = n^{-1}I$. Let the non-degeneracy Condition 20.2 be fulfilled. Then for any $\psi \in (0,1)$, $M < \infty$, $K < \infty$, $\varepsilon > 0$ there exists $\rho > 0$ such that for any convex in $\mathbb{R}^n$ simple cell complex $S \subset T$ of the class $C^3(M,K,\psi,\varepsilon)$,*

$$P(\max_{\mathbf{t} \in S} X(\mathbf{t}) > u) = \varphi(u) \sum_{v=0}^{n-1} \frac{\binom{n}{v} H_{n-1-v}(u)}{\omega_v (2\pi n)^{(n-v)/2}} W_v(K(\mathbf{T}))$$

$$+ \int_u^{\infty} \varphi(x)dx + O\left(\exp\left(-\frac{1}{2}u^2(1 + 1/n)\right)\right), \quad u \to \infty. \qquad (20.10)$$

Proof. By the comparison theorem (Theorem 20.8), we can assume without any loss of generality that the field X is homogeneous. Let us introduce a Gaussian field $Y(\mathbf{t})$ on $\mathbb{R}^n$ as follows. Denote

$$Y_1(\mathbf{t}) := \sum_{1 \leqslant k_1 < k_2 < k_3 \leqslant n} [X^0_{k_1 k_2 k_3} \cos(a(t_{k_1} + t_{k_2} + t_{k_3}))$$

$$+ Y^0_{k_1 k_2 k_3} \sin(a(t_{k_1} + t_{k_2} + t_{k_3}))];$$

$$Y_2(\mathbf{t}) := \sum_{1 \leqslant k_1 < k_2 \leqslant n} [X^1_{k_1 k_2} \cos(a(t_{k_1} + t_{k_2})) + Y^1_{k_1 k_2} \sin(a(t_{k_1} + t_{k_2}))$$

$$+ X^2_{k_1 k_2} \cos(2a(t_{k_1} + t_{k_2})) + Y^2_{k_1 k_2} \sin(2a(t_{k_1} + t_{k_2}))$$

$$+ X^3_{k_1 k_2} \cos(3a(t_{k_1} + t_{k_2})) + Y^3_{k_1 k_2} \sin(3a(t_{k_1} + t_{k_2}))];$$

$$Y_3(\mathbf{t}) := \sum_{1 \leqslant k_1 \leqslant n} \sum_{j=1}^{4} X^j_{k_1} \cos(jat_{k_1}) + Y^j_{k_1} \sin(jat_{k_1}).$$

Here the collection of all X's and Y's with upper and lower indexes are independent standard Normal variables. Denote further $a = \sqrt{N/n}$, with $N = \binom{n}{3} + 3\binom{n}{2} + 4n$ (the number of all summands in all three sums). The Gaussian field

$$Y(\mathbf{t}) = N^{-1/2}(Y_1(\mathbf{t}) + Y_2(\mathbf{t}) + Y_3(\mathbf{t}))$$

is the restriction of the field $Y_N(\mathbf{t})$, $\mathbf{t} \in \mathbb{R}^N$, to the linear subspace $\mathbb{L}_n \subset \mathbb{R}^N$, which is isomorphic to $\mathbb{R}^n$ and defined by the equations $t_1' = a(t_1 + t_2 + t_3), \ldots, t_N' = 4at_n$ (compare with the arguments in the above three sums defining $Y(\mathbf{t})$). The non-degeneracy Condition 20.2 is fulfilled for $Y(\mathbf{t})$, $\mathbf{t} \in \left[0, \pi\sqrt{n/N}/2\right]^n$. Now take $\tau \in (0, \pi\sqrt{n/N}/2)$ small enough so that the cube $\mathbf{b} + [0,\tau]^n \subset \mathbb{L}_n$ expressed in the coordinates of the manifold $\mathbb{L}_n$ is contained in $[0,\pi]^N \subset \mathbb{R}^N$ for some $\mathbf{b}$. Let us introduce the functional $\varphi(\cdot)$ defined on convex subsets $A \subset \mathbb{L}_n$ with $gA \subset [0,\tau]^n$ for all $g \in G_n$ by the relation

$$\varphi(A) = \int_{G_n} P(gA \cap B_u^1 \neq \varnothing)\omega(dg),$$

where $\omega(\cdot)$ is the Haar measure on G_n, normalized to one, i.e., $\omega(G_n) = 1$. Let us recall that $gA \subset [0,\tau]^n$ means that for some $\mathbf{b}$, $g_0A + \mathbf{b} \subset [0,\tau]^n$, g_0 is the corresponding rotation. Let us denote by $\mathcal{K}$ the set of all convex compacts in $\mathbb{R}^n$. By Theorem 4.2.1, Santalo [2004], φ is C-additive in $[0,\tau]^n$ in the sense that for any $A, B \in \mathcal{K}$, with $A \cup B \in \mathcal{K}$ and $A, B, A \cup B \subset [0,\tau]^n$, $\varphi(A \cup B) + \varphi(A \cap B) = \varphi(A) + \varphi(B)$. Now, using this property, we shall expand φ to all elements of $\mathcal{K}$. For $K, K' \in \mathcal{K}$ and $K, K' \subset [0,\tau]^n$, set

$$\varphi(K \cup K') = \varphi(K) + \varphi(K') - \varphi(K \cap K'),$$

and so on. Since any convex set can be partitioned into convex sets with arbitrary small diameters, using, for example, the coordinate grid, we thus have expanded our functional. Now we are in a position to show that our expansion is well defined. Let $C, C', K, K', C \cup C', K \cup K' \in \mathcal{K}$ be such that $C \cup C' = K \cup K'$ and $C, C', K, K' \subset [0,T]^n$ for some $T > 0$ and let C-additivity of φ hold for the convex compacts lying in $[0,T]^n$. Let us show that

$$\varphi(C) + \varphi(C') - \varphi(C \cap C') = \varphi(K) + \varphi(K') - \varphi(K \cap K').$$

By C-additivity,

$$\varphi(K) = \varphi(K \cap C) + \varphi(K \cap C') - \varphi(K \cap C \cap C'),$$

$$\varphi(K') = \varphi(K' \cap C) + \varphi(K' \cap C') - \varphi(K' \cap C \cap C'), \qquad (20.11)$$

$$-\varphi(K \cap K') = -\varphi(K \cap K' \cap C) - \varphi(K \cap K' \cap C') + \varphi(K \cap K' \cap C \cap C')$$

$$\varphi(C) = \varphi(C \cap K) + \varphi(C \cap K') - \varphi(C \cap K \cap K'),$$

$$\varphi(C') = \varphi(C' \cap K) + \varphi(C' \cap K') - \varphi(C' \cap K \cap K'), \qquad (20.12)$$

$$-\varphi(C \cap C') = -\varphi(C \cap C' \cap K) - \varphi(C \cap C' \cap K') + \varphi(C \cap C' \cap K \cap K')$$

and the sums of the right-hand sides of (20.11) and (20.12) coincide. Hence the sums of the left-hand sides coincide as well. Thus $\varphi(A)$, $A \in \mathcal{K}$, is well defined. It is invariant with respect to translations and rotations and, obviously, it is monotone with respect to the inclusion order. By Hadwiger's theorem, $\varphi = \sum_{k=0}^{n} a_k W_k$, $a_k \geqslant 0$. Now, let $A \in \mathcal{K}$ and $A \subset [0, \tau]^n$ satisfy the theorem conditions. Then

$$
\begin{aligned}
&\left| P(\max_{t \in A} Y(\mathbf{t}) > u) - \varphi(A) \right| \\
&= \left| \int_{G_n} \left(P(\max_{t \in A} Y(t) > u) - P(gA \cap B_u^1 \neq \varnothing) \right) \omega(dg) \right| \\
&\leqslant \int_{G_n} \left| P(\max_{t \in A} Y(t) > u) - P(\max_{t \in gA} Y(t) > u) \right| \omega(dg) \\
&\quad + \int_{G_n} \left| P(\max_{t \in gA} Y(t) > u) - P(gA \cap B_u^1 \neq \varnothing) \right| \omega(dg) \\
&\leqslant L \exp\left(-\frac{1}{2} u^2 (1 + \rho_2) \right),
\end{aligned}
\tag{20.13}
$$

where the first summand in the penultimate step is bounded by Corollary 20.11; the second one is bounded by Lemma 20.13. Moreover, $\rho_2 = \min(\rho, \rho_1)$. Furthermore, letting $A = \bigotimes_{i=1}^{n} [0, T_i]$ and substituting successively $T_2 = \cdots = T_n = 0$, $T_3 = \cdots = T_n = 0, \ldots$, we get the values of the coefficients a_k. Thus the statement of the theorem holds for all fields X and all convex closed sets satisfying the theorem conditions with sufficiently small diameters.

The nest step of the proof is to expand this result to sets with arbitrary diameter. Notice first that if $S \in C^3(M, K, \psi, \varepsilon)$, then

$$S_1 = S \cap \{t_1 \geqslant 0\}, \quad S_2 = S \cap \{t_1 \leqslant 0\}, \quad S_3 = S \cap \{t_1 = 0\}$$

are also of this class, perhaps with other M, K, ψ and ε. Therefore, by C-additivity of Minkowski functionals, Matheron [1976], it follows from the theorem assertion, which holds in this case for any $S \in \mathcal{K}$, $S \subset [0, \tau]^n$, that

$$P(u, S) = P(u, S_1) + P(u, S_2) - P(u, S_1 \cap S_2) + O(\exp(-u^2(1+\rho)/2), \tag{20.14}$$

as $u \to \infty$, where the positive ρ can differ from that in (20.10). Here $P(u, A)$ denotes the probability that the field is below u on A.

We again use the standard method for expanding excursion probabilities to greater parameter sets. If A, $B \subset T$ and $X(t)$ satisfies the conditions of the theorem, by Dmitrovsky's inequality,

$$P(\sup_{\mathbf{t}\in A} X(\mathbf{t}) > u, \sup_{\mathbf{t}\in B} X(\mathbf{t}) > u) \leqslant P(\sup_{\mathbf{s}\in A, \mathbf{t}\in B} X(\mathbf{s}) + X(\mathbf{t}) > 2u)$$

$$\leqslant P(\sup_{\mathbf{s}\in T, \mathbf{t}\in T} X(\mathbf{s}) + X(\mathbf{t}) > 2u) \leqslant L u^a \exp\left(-\frac{u^2}{1 + r_{AB}}\right), \quad (20.15)$$

where L, a do not depend of A, B, moreover, by non-degeneracy,

$$r_{AB} = \sup_{\mathbf{s}\in A, \mathbf{t}\in B} |r(\mathbf{t} - \mathbf{s})| < 1.$$

Denote $\tau_1 = \tau/n$. Any convex set with the diameter not exceeding τ_1 contains in $g[0,\tau]^n$ for any $g \in G_n$. Let S be a convex set satisfying the theorem conditions and belonging to the rectangle $K(\mathbf{T})$, $\mathbf{T} = (T_1,\ldots,T_n)$, $T_j \leqslant \tau_1, j = 1,\ldots,n-1, T_n > 0$. Let an integer k_n be such that $T_n/k_n \leqslant \tau_1/2$. Let us partition the interval $[0,T_n]$ into intervals with length T_n/k_n, and denote them by $\pi_1,\ldots,\pi_{k_n}$. Denote also $S_i = S \cap \pi_i, i = 1,\ldots,k_n$. By (20.14), (20.15) and homogeneity,

$$P(u,S) = \sum_{i=1}^{k_n} P(u,S_i) - \sum_{i=1}^{k_n-1} P(\max_{S_1} X(\mathbf{t}) > u, \max_{S_i} X(\mathbf{t}) > u)$$

$$+ O\left(\sum_{i>j+1} P(\max_{S_i} X(\mathbf{t}) > u, \max_{S_j} X(\mathbf{t}) > u)\right)$$

$$= \sum_{k=0}^{n} a_k(u) W_k(S) + O(\exp(-u^2(1 + \rho)/2)), \quad u \to \infty,$$

where positive ρ can be different from the one before. Then we consider sets in rectangles with an arbitrary long T_{n-1}, T_n and whose other edges are short, and repeat the same trick. On the last step we consider rectangles with arbitrary long $T_2,\ldots,T_{n-1}, T_n$ and repeat the same for T_1.

References

В. И. Питербарг. *Двадцать лекций о гауссовских процессах*. МЦНМО, Москва, 2015.

R. J. Adler. *The Geometry of Random Fields*. Wiley, London, 1981.

R. J. Adler and J. E. Taylor. *Random Fields and Geometry*. Springer Monographs in Mathematics. Springer, 2007.

L. B. Andersen and V. V. Piterbarg. *Interest Rate Modeling, in Three Volumes*. Atlantic Financial Press, 2010.

T. W. Anderson. *An Introduction to Multivariate Statistical Analysis*. Wiley, 2003.

J.-M. Azaïs and M. Wschebor. *Level sets and extrema of random processes and fields*. Wiley, 2009.

P. Billingsley. *Convergence of Probability Measures*. Wiley, 1999.

S. Chatterjee. An error bound in the Sudakov-Fernique inequality. ArXiv working paper, February 2008.

R. M. Dudley. The sizes of compact subsets of Hilbert space and continuity of Gaussian processes. *J. Funct. Anal.*, 1:290–330, 1967.

A. I. Elizarov. On the variance of the number of stationary points of a homogeneous Gaussian field. *Theory of Probability and Its Applications*, 29(3): 569–570, 1985.

X. Fernique. Regularité des trajectoires des fonctions aléatoires gaussiennes. *Lecture Notes in Mathematics*, 480, 1975.

D. Geman. On the variance of the number of zeros of a stationary Gaussian process. *Annals Math. Stat*, 43:977–982, 1972.

H. Hadwiger. *Vorlesungen Über Inhalt Oberfläche und Isoperimetrie*. Springer-Verlag, Berlin-Göttingen-Heidelberg, 1957.

I. A. Ibragimov and Y. A. Rozanov. Gaussian random processes. *Applications of Math. Springer-Verlag, New York-Heidelberg-Berlin*, 9, 1978.

K. Ito, K.-I. Sato, and O. E. Barndorff-Nielsen. *Stochastic Processes*. Springer, 2004.

O. Kallenberg. *Random Measures*. Academic Press, New York, London; Akademie-Verlag, Berlin,, 4th edition edition, 1986.

J. Kim and D. Pollard. Cube root asymptotics. *Ann. Statist.*, 18(1):191–219, 1990.

A. N. Kolmogorov and S. V. Fomin. *Elements of the Theory of Functions and Functional Analysis*. Dover Books on Mathematics, 1999.

V. P. Leonov. On variance of time average for a stationary random process. *Theory of Probability and its Applications*, 1:93–101, 1961.

G. Matheron. *Random Sets and Integral Geometry*. J. Wiley&Sons, New-York-London-Sydney-Toronto, 1976.

V. I. Piterbarg. Comparison of distribution functions of maxima of Gaussian processes. *Theory of Probability and its Applications*, 26(4):68–706, 1982.

V. I. Piterbarg. High deviations for multidimensional stationary Gaussian processes with independent components. In V. M. Zolotarev, editor, *Stability Problems for Stochastic Models*, pages 197–230. Springer, 1994.

V. I. Piterbarg. *Asymptotic Methods in the Theory of Gaussian Processes and Fields*. American Mathematical Society, 1996.

S. O. Rice. Mathematical analysis of random noise. *Bell System Tech. J.*, 23: 282–332, 1944.

Y. Rozanov. *Stationary Random Processes*. Holden-Day, San Francisco, 1967.

L. A. Santalo. *Integral Geometry and Geometric Probability*. Cambridge Mathematical Library. Cambridge Mathematical Library, 2004.

S. Willard. *General Topology*. Addison-Wesley, Reading Massachusetts, 1970.

www.ingramcontent.com/pod-product-compliance
Lightning Source LLC
Chambersburg PA
CBHW070724220326
41598CB00024BA/3288